Dances from Norway

by Daniel Beal

Minneapolis

Printed in the United States of America
Precision Graphics Inc., Plymouth, Minnesota
1988

For information, contact: Sons of Norway
1455 West Lake Street, Minneapolis, MN 55408

International Standard Book Number: 0-9620940-0-5
Library of Congress Card Number: 88-092216

FRONT COVER: *Daniel and Gaelyn Beal.*
Photo by Clyde Kumpula.

BACK COVER: *Norwegian folk dancers (Symra Leikarringen).*
Drawing by Lary Anderson.

Contents

Index

Children's Dances

Song-dances

Figure-dances

Old-time Couple Dances

V. Easy	Very Easy
Int.	Intermediate
Adv.	Advanced

Acknowledgements

I would like to express my sincere appreciation to all who have helped me with some aspect of this book: Lynn Dahlquist Moluf for taking photos of all the dance steps and positions; Jack Norman for rendering those photos into drawings; Sogn og Fjordaneringen and Symra Leikarringen for giving me the opportunity to photograph their groups in action; Noregs Ungdomslag and the director of its dance council, Dag Vårdal, for reviewing the song-dances and figure-dances; Gaelyn Beal and Robert Moluf for editing the historical notes; Liv Nordem Lyons for editing my translations of Norwegian songs; Ånon Egeland for finding melodies for several of the figure-dances; Roger Tweiten for notating many melodies from my field recordings; LeRoy Larson for allowing me to use his notation of Ellingson's polka; John Schubert for helping me review the music engravings; and Claudette Sexton for working through the difficulties in typesetting a book of this nature.

Special thanks go to the dancers in the Røros area and to the members of the Symra Leikarringen in Oslo, who taught me new dances for this volume. I am particularly grateful to Erik Hestnes of the Symra Leikarringen, who offered his generous assistance in a variety of ways, from arranging photo sessions in Norway to editing most of the dance directions over the course of several years.

Funding for this book was provided in part by a grant from the Sons of Norway Foundation as a part of its cultural programs. I would like to thank them, as well as Mr. and Mrs. E. Maynard Beal, who contributed additional funding.

Previously published song-dance and figure-dance melodies were newly engraved and given introductions for *Dances from Norway.* The melodies are presented with the permission of Noregs Boklag.

The new English descriptions of song-dances and figure-dances were created in cooperation with Noregs Ungdomslag and are presented with the permission of that organization.

Introduction

I haved enjoyed teaching Norwegian dances in the United States and Canada over the last twelve years. During that time, I have received many requests for a book that would give instructions in English for a variety of Norwegian folk dances and tell something of their history. *Dances from Norway* was written in response to those requests.

Norwegian dances may be divided into four groups: song-dances, figure-dances, old-time couple dances, and provincial dances (the Norwegian names are songdanser, turdanser, gammeldanser, and bygdedanser). For this book, I have selected some of the more popular dances from the first three groups. I have also included one Norwegian-American dance, Seattle Polka, and a few Norwegian dances for children. Provincial dances were omitted because of their complexity.

Somewhat like a recipe, the first part of this book lists all of the basic ingredients that go into the making of the dances in this collection. Explanations for such things as common handholds and dance steps may be found here, together with many illustrations. The next part of the book gives historical notes and directions for each group of dances—song-dances, figure-dances, and old-time couple dances.

The historical notes involved considerable detective work. Besides conducting research in Norwegian libraries, I sifted through dozens of travelogs in the special collections of the University of Minnesota, the University of London, and the British Library with the hope of uncovering some early accounts of dancing in Norway. A few of my finds are quoted in the text. I have also looked to sources that consider dances from other European countries. More so than any previous author, I have attempted to view Norwegian dances in light of trends that were popular throughout Europe.

Many of the dances in this book were first written down by a Norwegian author, Klara Semb, in her book, *Norske Folkedansar.* While *Norske Folkedansar* has been a standard reference for many years, Semb's instructions are not always very clear, especially for those who may not have had much experience with Norwegian dances. I have attempted to look at the dances from a fresh perspective and provide more detailed instructions.

My instructions for the song-dances and figure-dances were reviewed with the assistance of Noregs Ungdomslag (Norwegian Youth League). Among other undertakings, this organization encompasses the largest folk-dance society in Norway, sponsoring workshops and groups across the country.

Music for all of the dances may be found in the last part of the book. Song-dance melodies include verses with English translations.

—Daniel Beal
Minneapolis, 1988

How To Use This Book

First

Take a moment to become acquainted with the few abbreviations and special terms that are found on pages 7-9. These will be used frequently in the dance directions.

Second

Choose a dance. The level of difficulty may be found in the index on pages 1-2.

If you choose a song-dance, it is advisable to learn at least the first verse of the accompanying song before attempting to do the dance. You can then provide your own music as you go along. To show how the first verse fits together with the various steps and actions, the directions indicate which words should be sung for each part of the dance.

Third

Read the directions. Handholds, positions, and step patterns may be looked up as needed.

Step patterns are explained with a series of drawings. Example:

May begin on either foot
Each step pattern alternates feet
Music 4/4

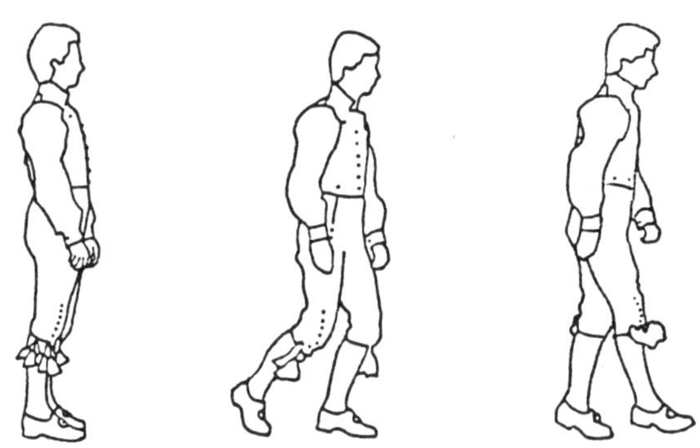

starting position	1 step L	2 hop l

Many step patterns may begin on either foot. This is noted in the upper left corner.

The musical counts and footwork are given below each drawing. L stands for left foot, R for right foot, and B for both feet. *A small L or R means that your weight remains on the supporting foot during the next action.*

Many of the steps should be performed with a small dip in which you bend and extend your ankle and knee. This dipping action is indicated by a bent arrow:

1
step
↘↗
L

In other cases, you should descend on one step and rise on the next. This is indicated by two arrows, one pointing downward, one upward:

1
step
↘
L

2
close
↗
R

Some patterns include a hop. The upward spring of a hop is represented by a black dot. Since this upward spring is immediately followed by a dip, a bent arrow appears after the dot. •↘↗

Don't forget to look at the arrows. A step pattern is bound to feel stiff and awkward unless you remember to bend and extend your ankles and knees at the right times.

Diagrams

Man as seen from above. The "nose" indicates the direction he is facing.

Woman as seen from above. The "nose" indicates the direction she is facing.

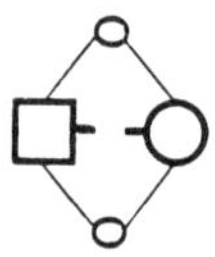

Man and woman as seen from above. Arms are extended to either side; small circles represent hands.

Glossary

Abbreviations

CCW	Counterclockwise
Ct	Count
CW	Clockwise
LOD	Line of Direction
Ms	Measure (as used in music)

Dance Terms

Acknowledgement: A brief recognition of the dancer across from you (usually partner) with a slight bow of the head and upper body.

Chain: A figure in which men and women weave past each other in a circle formation (also known as a "grand-right-and-left"). To perform a chain, partners face each other in a circle formation:

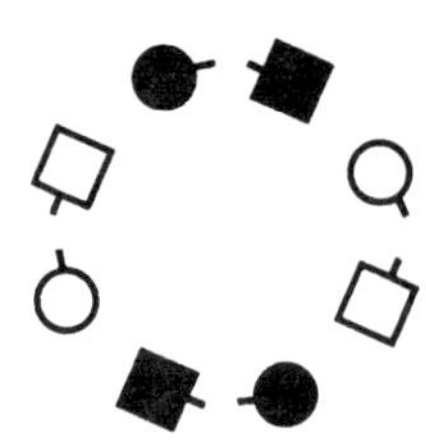

Take right hands with your partner at about shoulder level. Curve to the left of your partner, such that you pass right shoulders:

As you go past your partner, release right hands and offer your left hand to the next person of the opposite sex. Curve to the right of this person, such that you pass left shoulders:

The chain continues in the same way with the next dancers you meet. Each dancer you pass in the chain is counted as one person. If you have gone around the circle and are about to meet the same dancers again, continue to count off each person with a consecutive number.

Circle A: Several couples form a circle. Man is to left of partner.

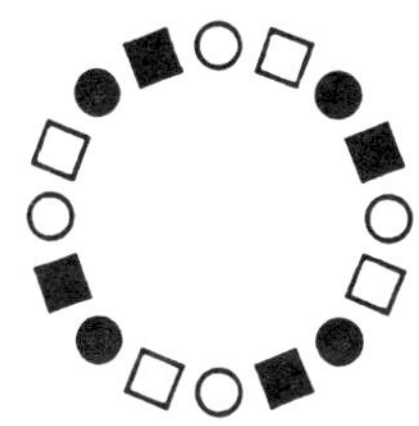

Circle B: Several couples form a circle, such that they are positioned like spokes of a wheel. Man is closest to center of circle.

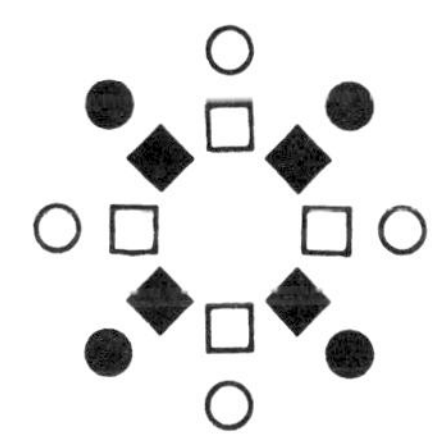

Close: An action where the free foot comes to place beside the supporting foot. Unless stated otherwise, all of the dancer's weight will be transferred to the free foot.

Corner person: A person of the opposite sex who is generally on the other side of you from your partner. Corner persons for the man and woman in black are shown below:

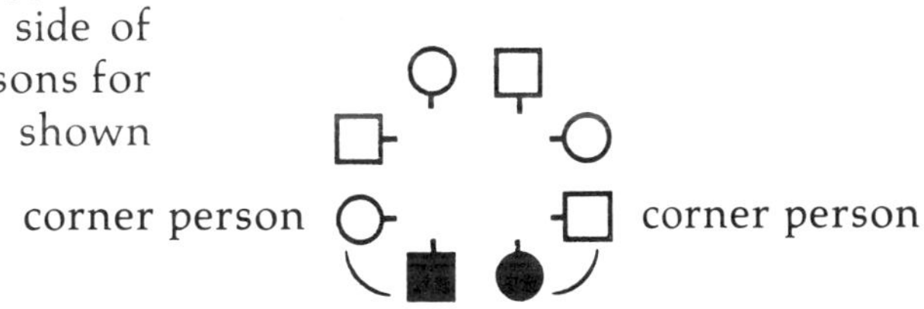

Line of direction: The pathway or course in which the action of a dance takes place. The line of direction may also serve as an orientation for a dancer's position in a circle formation:

facing CCW LOD (counterclockwise line of direction)

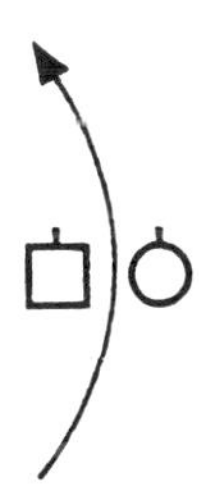

facing CW LOD (clockwise line of direction)

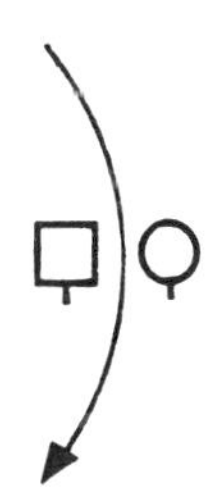

Phrase: A brief melody, consisting of several measures, that makes up one part of a larger piece of music. In the dance directions, the beginning of each new phrase is indicated in the left-hand margin with a capital letter A, B, C, etc. Each part of a Norwegian dance is usually done to one phrase of music.

Touch: An action where the free foot lightly touches the floor. *Weight is not transferred to that foot.*

Handholds

Closed Handhold

Several couples form a circle. Everyone faces CW LOD with each man in front of his partner.

With his right hand, man takes left hand of his partner and tucks her forearm beneath his arm, somewhat like one might do with a cane. Thumbs should be uppermost and palms are together. Man extends his left arm forward to woman who is in front of him.

In a like manner, woman takes left hand of man behind her with her right hand, while extending her left forward to her partner.

Everyone should be fairly close together with upper bodies angled a little toward center of circle.

Light Handhold

Stand side-by-side with partner. Depending on dance, man may be to left or right of partner.

Join inside hands at about shoulder level. Man holds woman's hand from beneath.

Light Handhold Extended to Side

Face your partner. Extend one arm to side and join hands with partner. Man holds woman's hand from beneath.

Depending on dance, man may take woman's left hand with his right or vice versa.

Light Handhold in Circle

Several couples form a circle.

Join hands at about shoulder level. Man holds woman's hands from beneath.

When traveling in a circle, everyone should generally face proper direction, but upper bodies are angled a little toward center of circle.

Low Handhold

Several couples form a circle. Everyone faces center of circle with man to left of his partner.

Join hands. Arms should be relaxed.

Open Handhold

An easy way to learn open handhold is to assume closed handhold first.

Without releasing hands, turn to face center of circle. Raise hands a little above shoulder level, such that your arms are bent in front of you—not to either side. As you dance, your arms should be relatively steady.

Hand on Hip/Skirt

Hand low on hip, fingers forward. (Thumb points backward).

Hand on hip, thumb forward. Palm of hand is pressed against waist.

Hands on hips, fingers forward. (Thumbs point backward.)

Woman holds skirt to either side.

Woman holds skirt to side.

Formal Greeting

Man:

Place left hand low on hip, fingers forward. Your right arm should be relaxed by your side.

Bow forward from the waist. At same time, slowly lift right arm in front of body with a graceful sweeping motion. Rise to a natural standing position. Your right arm should slowly return to place by your side.

Woman:

Place your weight on left foot. Hold skirt to either side.

Sweep right foot behind your left, tracing a half circle with tip of shoe.

With weight mostly on left foot, lower to a crouching position. Bow your head and tilt upper body a little forward. At same time, keep your back relatively straight.

To rise, shift your weight backward to right foot. Your left foot should remain in place and extend. Come to a standing position gracefully, now shifting your weight forward to left foot and closing right foot beside it. Lift your head at last moment.

Specific counts for formal bow and curtsey are given with dance directions.

Informal Greeting

Man:

Fold arms above chest and parallel to ground.

Bow forward from waist. Keeping arms folded, rise to a natural standing position.

Woman:

Place your weight on right foot. Hold skirt to either side.

Touch ball of left foot several inches behind your right.

With weight still on right foot, bow your head and lower to a kneeling position, using left foot for balance. Your left knee should touch floor approximately by instep of right foot. Keep your back relatively straight and let your shoulders round a little forward.

Rise with your weight on right foot. Come to a standing position gracefully, closing left foot beside your right. Lift your head at last moment.

Specific counts for the informal bow and curtsey are given with the dance directions.

Positions

Position 1

Man places right hand around woman's waist. He extends left arm out to side (somewhat bent) and grasps woman's right hand with his left from above.

Woman places left hand on man's shoulder. She extends right arm out to side (somewhat bent) with palm of hand facing downward.

Position 2

Man places both hands around woman's waist.

Woman places both hands on man's shoulders. Each hand should be centered on shoulder.

Note: When performing old-time couple dances, Norwegian men often place both hands around woman's shoulder-blade area instead of waist.

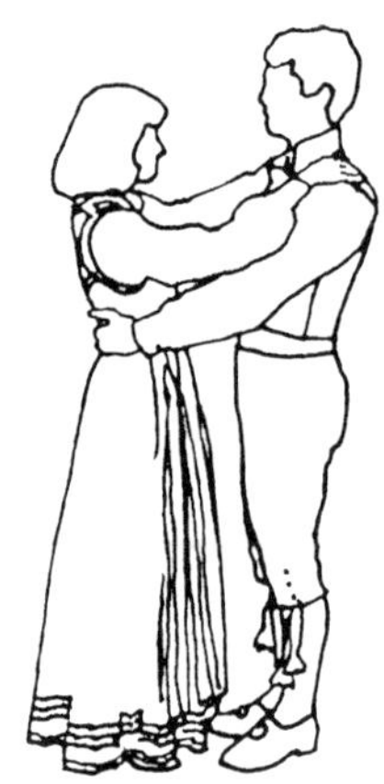

Position 3

Face your partner. Place your left hand on upper right arm of your partner. Place your right hand around his or her shoulder-blade area. Elbows should be lifted.

Position 4

Stand almost side-by-side with your partner. Place your left hand on upper right arm of your partner. Place your right hand around his or her waist. Your left elbow should be lifted a little.

It is also possible to face your partner with this position, but this variation is not used here.

Position 5

Stand side-by-side with partner. Extend your right arm behind your partner's back, such that arms cross. Bend your left arm and take your partner's right hand with your left.

Position 6

Face your partner. Extend your arms to either side (somewhat bent). Palms of hands should be facing downward.

Man grasps woman's hands from above, curving around outer edges of her hands.

Step Patterns

Arm-swing Turn

May begin on either foot
Each step pattern alternates feet
Music 2/4

First arm-swing turn

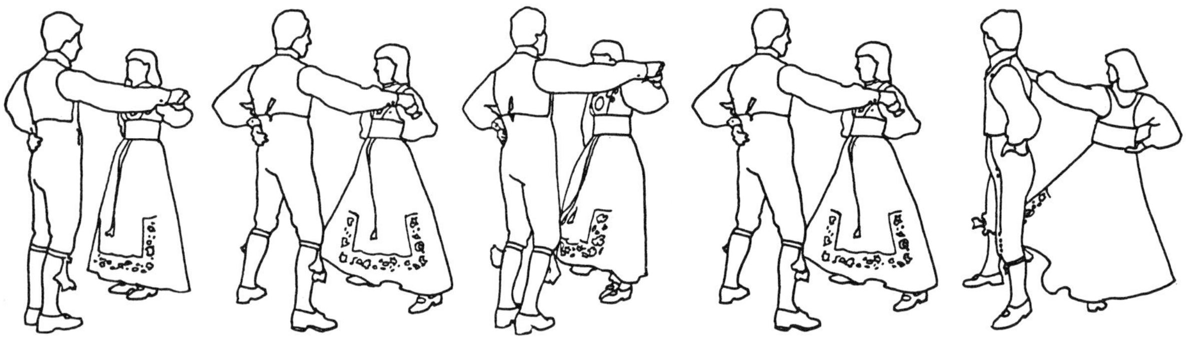

	starting position	1 step ↘	& close ↗	2 step ↘	& (pivot) ↗
man:		L	R	L	l
woman:		R	L	R	r

Second arm-swing turn

	1 step ↘	& close ↗	2 step ↘	& (pivot) ↗
man:	R	L	R	r
woman:	L	R	L	l

NOTES:

Partners face each other with light handhold extended to side (man's right hand, woman's left). Free hand is on hip, thumb forward.

Perform first arm-swing turn face-to-face. At end of pattern, turn away from your partner. Your joined hands should sweep down and forward. Perform second arm-swing turn nearly back-to-back. At end of pattern, turn toward your partner. Your joined hands should sweep down and backward.

An arm-swing turn may be done in opposite direction with opposite footwork.

Change Step

May begin on either foot
Each step pattern alternates feet
Music 2/4

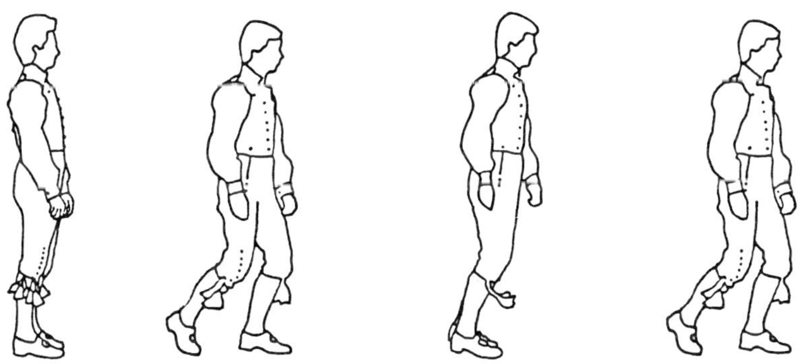

starting position	1 step ↘ L	& close ↗ R	2 step ↘↗ L

Music 3/4: First version

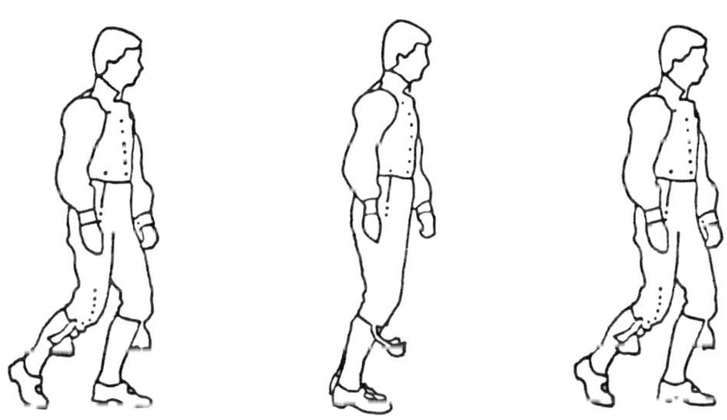

1 step ↘ L	2 close ↗ R	3 step ↘↗ L

Music 3/4: Second version

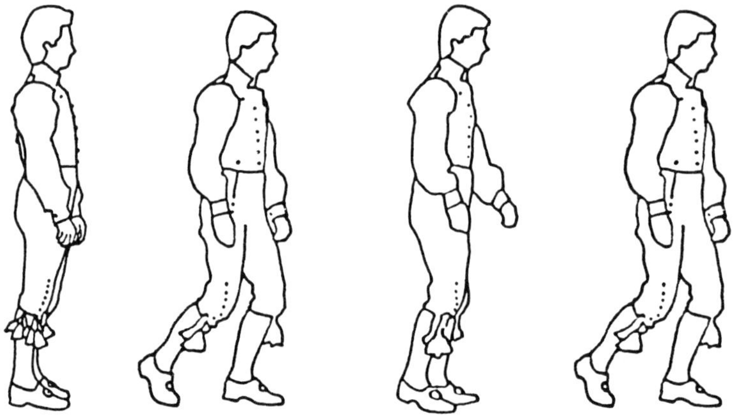

starting position | 1 step L | 2 close R | 3 step L

NOTES:

Notice that bending and extending action differs in last step pattern.

After last step, swing free foot forward in a natural manner to begin next pattern.

Closed Ballad Step

Each step pattern begins on left foot
Music 3/4*

starting position | 1 step L | 2 close R | 3 step L

1	2	3
(dip)	step	(dip)
l	R	r

NOTES:

*May also be counted 1-2-3-4-1-2 in 4/4 rhythm or 1-&-2-&-1-& in 2/4 rhythm.

A closed ballad step has a total of 6 cts. It may begin or end in middle of a musical ms.

Each dip is performed in place on supporting foot. At same time, touch ball of free foot beside supporting foot.

Steps are light and buoyant, springing from ball of foot.

Directions indicate which words should be sung for each part of dance. Below words you will find footwork for each closed ballad step. Counts are given where these might be helpful. Example:

Eg	rod-de meg	ut	på	sei-e	grun-	nen
	1	2	3	1	2	3
	L	R	L	l	R	r

Extending Step

May begin on either foot
Each step pattern alternates feet
Music 2/4

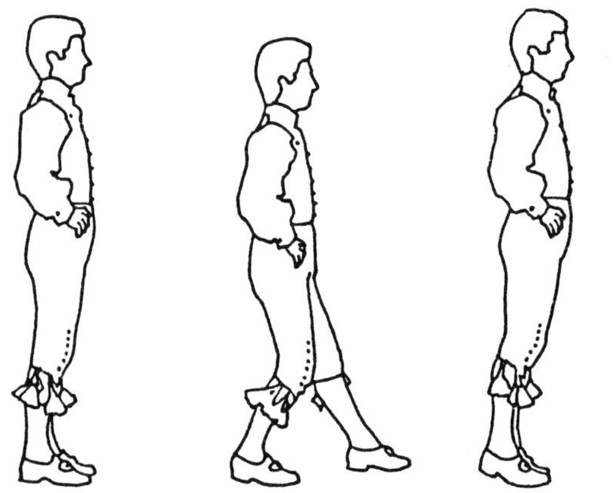

starting position — 1 step ↘ R — & (close left foot)

NOTES:

An extending step is performed in place.

Lift off from ball of left foot with a small upward swing in preparation for first step.

On first ct, land with your weight on ball of right foot and descend, bending at ankle and knee. At same time, extend left foot forward. Your left leg should be held straight.

On "&" ct, close left foot beside right. As left foot comes to place, lift off from ball of right foot in preparation for next extending step.

Next extending step will be done with opposite footwork. Transfer weight smoothly to ball of left foot and extend right foot forward.

Gesture Step

May begin on either foot
Each step pattern alternates feet
Music 2/4

One gesture step to right; back view.

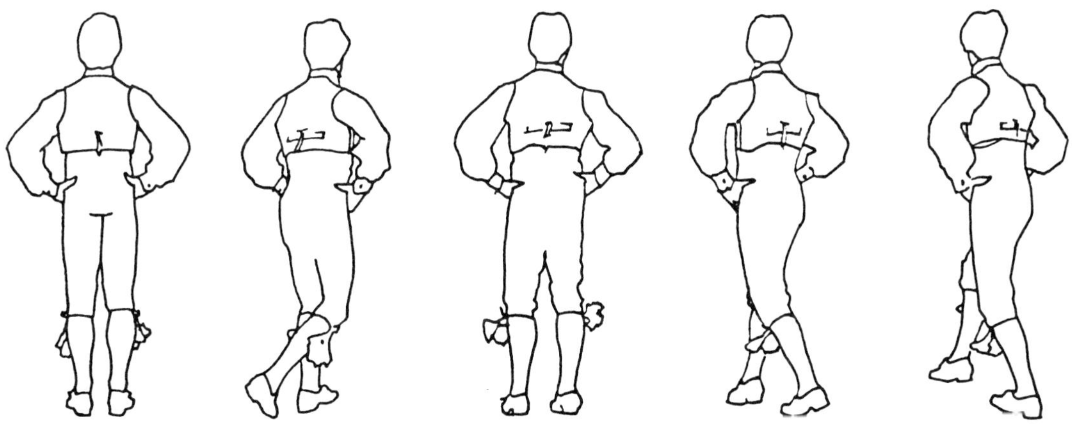

starting position	1	&	2	&
	cross	side	behind	hop
	↘	↗	↘	•↘↗
	L	R	L	l

NOTES:

A gesture step is performed almost entirely in place with some turning of body.

Hopping action is small. Ball of foot remains in contact with floor.

As you hop, lift your free leg from knee as if a string pulled it from above. Your foot is relaxed and points downward. It should be raised a short distance from floor.

After performing one gesture step to right, perform next gesture step to left with opposite footwork. Begin on right foot.

Hoppvals Step

May begin on either foot
Each step pattern alternates feet
Music 2/4

starting position	1	&	2
	step	close	step
	→	→	↘↗
	L	R	L

NOTES:

Step onto whole foot on first ct. Your leg should be comfortably bent.

On ct "&," man quickly closes ball of free foot near heel of other foot. Woman quickly closes ball of free foot a little in front of other foot.

Step forward/backward on last ct and perform a small dip.

Perform each step in a smooth and controlled manner.

Hoppvals steps are performed somewhat differently when partners turn together. Shift your weight a little more forward and toward balls of feet. On ct "&," quickly close free free foot toward instep of other foot. It should fall a little short of coming beside instep. As when turning with a reinlender step, you must turn a little with each hoppvals step in order to come completely around.

Light Running Step

May begin on either foot
Each step alternates feet

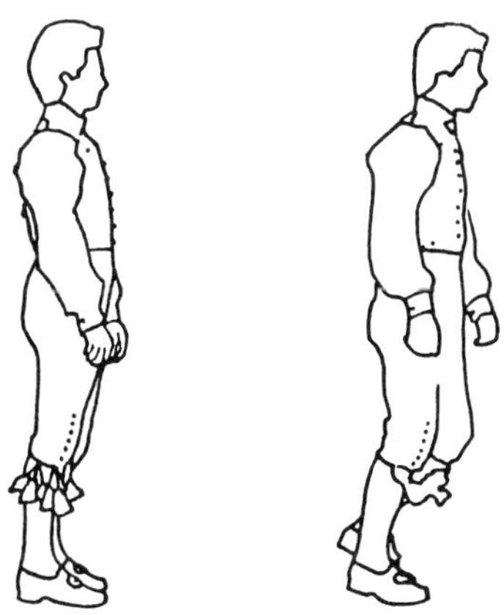

starting position

1
run
R

NOTES:

There is one step per beat.

Come down on ball of your foot first. Your foot should absorb weight in a natural manner, such that your heel will lower and rise a little from floor.

When performing a series of light running steps, footwork should be small and close to ground.

Light Step

May begin on either foot
Each step alternates feet

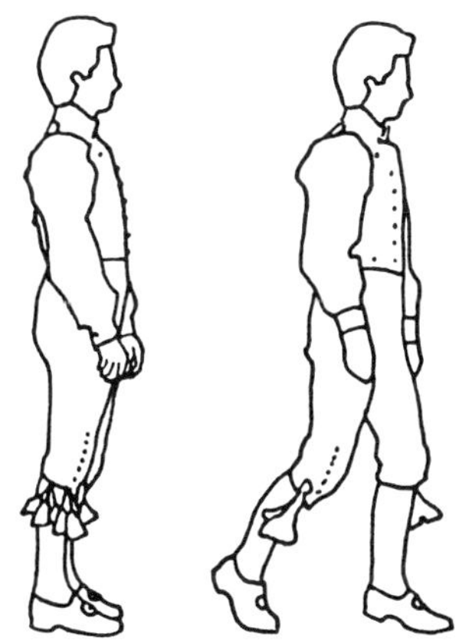

starting position

1
step
L

NOTES:

There is one step per beat.

Come down on ball of your foot first. Your foot should absorb weight in a natural manner, such that your heel will lower and rise a little from floor.

Lilting Step

May begin on either foot
Each step pattern alternates feet
Music 3/4

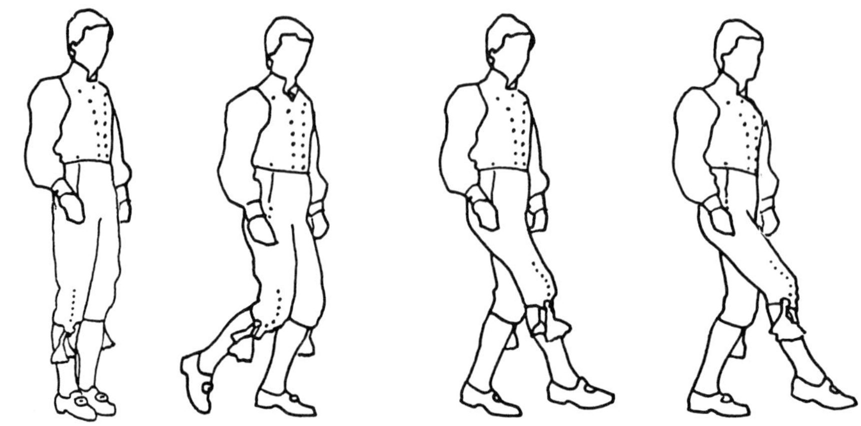

starting position	1 step	2 (rise)	3 (dip)
	↘	↗	↘↗
	L	l	l

NOTES:

Your weight will remain on 1 foot for all 3 cts. Step forward on first ct. Descend with this step, bending at ankle and knee. Rise on second ct. Your heel should lift a little from floor. Perform a small dip on third ct. This dip should serve as a natural weight-absorbing reflex.

As you rise and dip, swing your free leg forward, lifting a little from knee. Relax foot, such that it points downward. Your legs should be parallel to each other. There is no outward rotation from hip joints.

Perform next lilting step with opposite footwork.

Open Ballad Step

Each step pattern begins on left foot
Music 2/4*

starting position

1
step
L

&
close
R

2
step
L

&
(dip)
l

1
step
R

&
(dip)
r

NOTES:

*May also be counted 1-2-3-4-1-2 in 4/4 rhythm or 1-2-3-1-2-3 in 3/4 rhythm.

An open ballad step has a total of 6 cts. It may begin or end in middle of a musical ms.

Each dip is performed in place on supporting foot. At same time, let your free leg hang in a natural manner, such that your free foot lightly touches floor.

Each step should be light and buoyant. Your weight should be forward and toward balls of feet.

It is customary to look at your partner during first dip and at your corner person during second.

Directions indicate which words should be sung for each part of dance. Below words you will find footwork for each open ballad step. Counts are given where these might be helpful. Example:

Kva	er	det	vel_____		dei	sy-ta	for_____		i
&	1		2	3		1	2	3	
	L		R	L		l	R	r	

Reel Step

May begin on either foot
Each step pattern alternates feet
Music 2/4

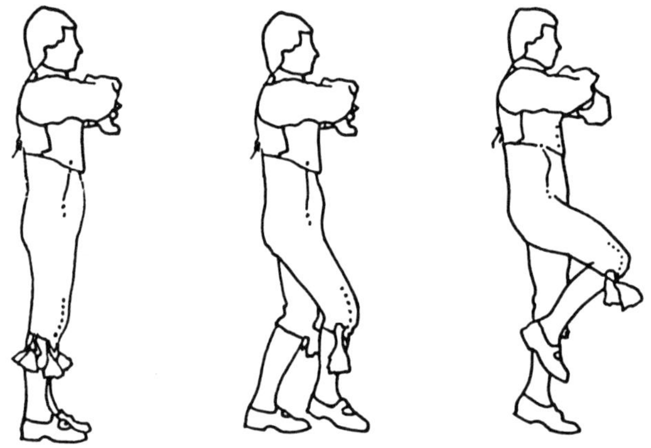

starting position	1	&
	step	scoot
	↘	•↗
	L	l

NOTES:

On the first ct, place ball of your left foot directly behind heel of your right foot.

On "&" ct, spring upward on your left foot, and at same time, scoot forward to where right foot was. Your right foot should lift upward and behind left foot in preparation for next reel step.

Reinlender Step

May begin either foot
Each step pattern alternates feet
Music 4/4*

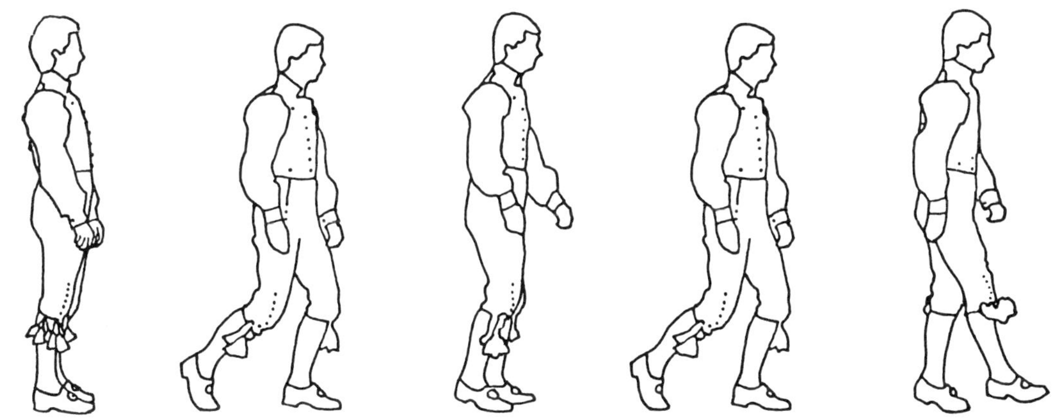

starting position	1 step L	2 close R	3 step L	4 hop l

NOTES:

*May also be counted 1 & 2 & in 2/4 rhythm.

Hopping action is small. Ball of supporting foot remains in contact with floor.

As you hop, swing free foot forward to begin next reinlender step.

Reinlender Step: Turning Variation

May begin on either foot
Each step pattern alternates feet
Music 4/4*

First reinlender step

	starting position	1 step	2 close	3 step	4 hop
		↘↗	↘↗	↘↗	•↘↗
man:		L	R	L	l
woman:		R	L	R	r

Second reinlender step

	1 step	2 close	3 step	4 hop
	↘↗	↘↗	↘↗	•↘↗
man:	R	L	R	r
woman:	L	R	L	l

NOTES:

*May also be counted 1-&-2-& in 2/4 rhythm.

Partners assume position 2 or 3, depending on dance. Man faces away from center of circle, woman toward:

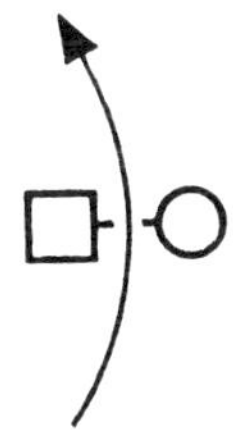

Partners make ½ CW turn with each reinlender step. At same time, they travel in CCW LOD.

You should turn a little CW with each step in order to come completely around. Note that each step is performed with a small dip.

Hopping action is small. Ball of foot remains in contact with floor. Use hopping action to continue turning, such that you pivot on ball of foot.

It is possible to turn CCW with reinlender steps. To do this, man begins by facing CCW LOD, woman CW LOD. Footwork is opposite of above, and partners turn in opposite direction.

Ringlender Step

Each step pattern begins on same foot
Music 4/4

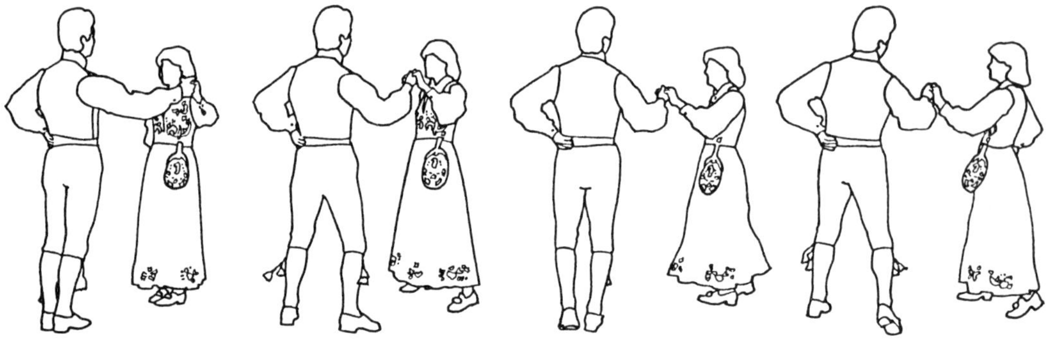

	starting position	1 step	2 close	3 step
man:		L	R	L
woman:		R	L	R

	4 hop	1 step	2 hop
man:	l	R	r
woman:	r	L	l

NOTES:

A ringlender step has a total of 6 cts. It may begin or end in middle of a musical ms.

Partners face CCW LOD and join inside hands with light handhold. Free hand is on hip, thumb forward.

Travel in CCW LOD with first part of this pattern (step, close, step-hop). Following this, turn a little toward your partner and perform last step-hop in other direction (CW LOD). Your joined hands should swing somewhat to side in a natural manner as you turn toward your partner.

Hopping action is small. Ball of your supporting foot remains in contact with floor.

To begin next ringlender step, turn to face CCW LOD, swinging joined hands forward.

Springpolka Step

Each step pattern begins on same foot
Music 3/4

	starting position	1 step	2 hop	3 step
man:		L	l	R
woman:		R	r	L

NOTES:

Partners assume position 2.

Man faces away from center of circle, woman toward:

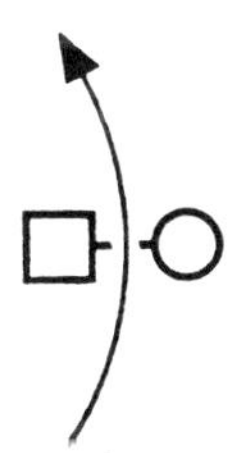

Partners make 1 CW turn with each springpolka step. At same time, they travel in CCW LOD.

On first ct, man turns onto his left foot, such that he faces CW LOD. At same time, woman takes 1 step forward and between man's feet with her right foot, such that she faces CCW LOD.

On second ct, continue to turn with hopping action.

On third ct, complete turn with 1 more step.

Step-Behind

Each step pattern begins on same foot
Music 2/4

starting position	1 step L	& behind R

NOTES:

A step-behind may be done in opposite direction with opposite footwork.

Step-Hop

May begin on either foot
Each step alternates feet
Music 2/4*

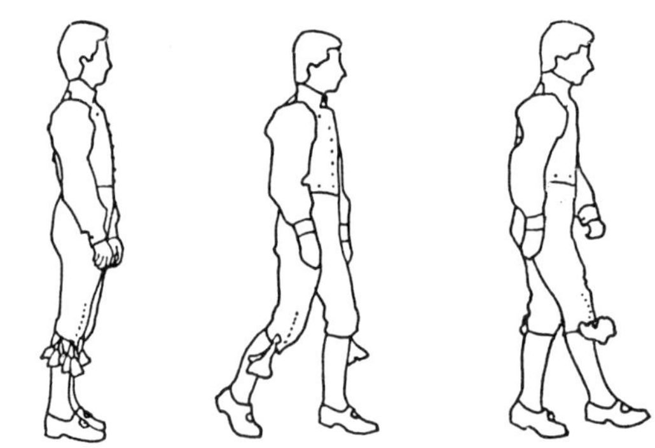

starting position	1 step L	& hop • l

Music 3/4

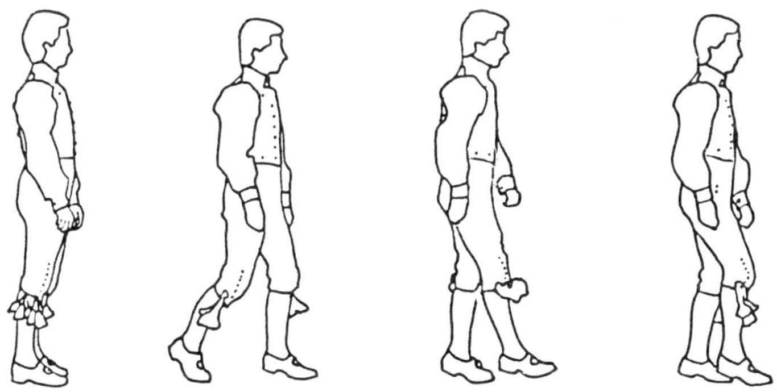

starting position	1	2	3
	step	hop	(dip)
	↘	•	∨↗
	L	l	l

NOTES:

*May also be counted 1-2 in 4/4 rhythm.

As you hop, swing free foot forward in a natural manner to begin next step-hop.

Step-Swing

May begin on either foot
Each step pattern alternates feet
Music 3/4

starting position	1	2	3
	step	(rise)	(dip)
	↘	↗	∨↗
	L	l	l

NOTES:

A step-swing is performed in place.

Your weight will remain on 1 foot for all 3 cts. Step in place on first ct. Descend with this step, bending at ankle and knee. Rise on second ct. Your heel should lift a little from floor. Perform a small dip on third ct. This dip should serve as a natural weight-absorbing reflex.

As you rise and dip, swing your right leg in front of your left. Relax foot, such that it points downward.

Perform next step-swing with opposite footwork.

Touch Step

May begin on either foot
Each step pattern alternates feet
Music 3/4

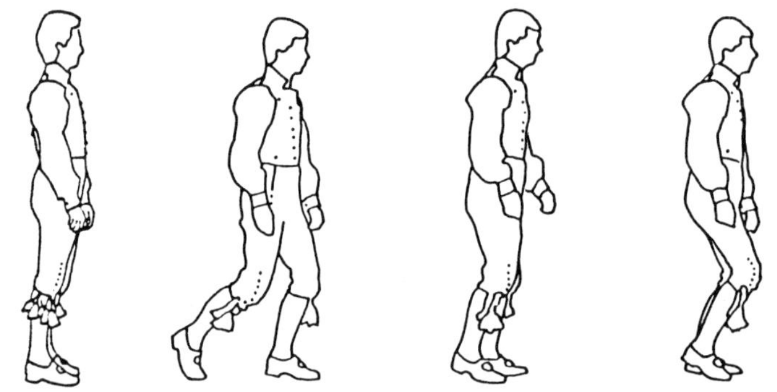

starting position	1 step	2 (rise)	3 (dip)
	L	l	l

NOTES:

Your weight will remain on 1 foot for all 3 cts. Step forward on first ct. Descend with this step, bending at ankle and knee. Rise on second ct. Your heel should lift a little from floor. Perform a small dip on third ct. This dip should serve as a natural weight-absorbing reflex.

As you rise and dip, touch ball of your free foot beside instep of your supporting foot.

Perform next touch step with opposite footwork.

Trio Step

May begin on either foot
Each step pattern alternates feet
Music 4/4

First trio step

	starting position	1 step	& close	2 step
		↘	↗	∨↗
man:		L	R	L
woman:		R	L	R

Second trio step

	3 step	& close	4 step
	↘	↗	∨↗
man:	R	L	R
woman:	L	R	L

NOTES:

Partners assume position 1, 2, or 3, depending on dance. Man faces away from center of circle, woman toward:

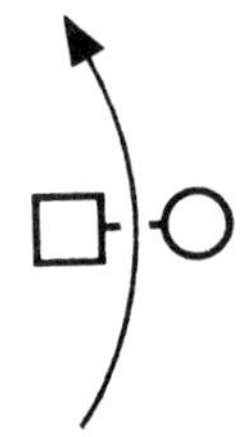

Partners make ½ turn with each trio step. At same time, they travel in CCW LOD.

On first ct, man turns onto his left foot, such that he faces CW LOD. At same time, woman takes 1 step forward and between man's feet with her right foot, such that she faces CCW LOD.

On second and third cts, continue to turn with 2 small steps.

Footwork is performed quickly.

Reverse roles for second trio step. Woman begins by turning onto her left foot. Man steps forward between woman's feet.

Waltz

May begin either foot
Each step pattern alternates feet
Music 3/4

First waltz step

	starting position	1 step	2 step	3 step
man:		L	R	L
woman:		R	L	R

Second waltz step

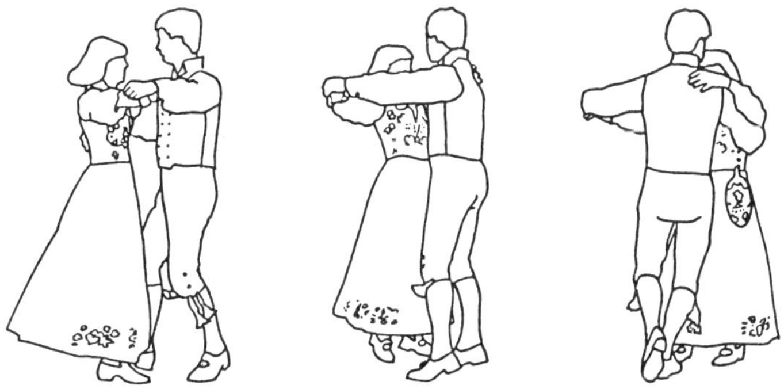

	1	2	3
	step	step	step
	↘	↗	↘↗
man:	R	L	R
woman:	L	R	L

NOTES:

Partners assume position 1 or 2, depending on dance.

Man faces away from center of circle, woman toward:

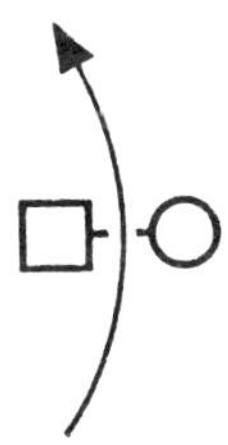

Partners make ½ CW turn with each waltz step. At same time, they travel in CCW LOD.

On first ct, man turns onto his left foot, such that he faces CW LOD. At same time, woman takes 1 step forward and between man's feet with her right foot, such that she faces CCW LOD.

On second and third cts, continue to turn with 2 small steps. Man performs these steps mostly in place. Woman dances a little forward, such that her steps curve in CW direction.

Note upward and downward motion of waltz step. Descend with first step, bending at knee and ankle. Rise on second step. Your heel should lift a little from floor. Perform a small dip with third step. This dip should serve as a natural weight-absorbing reflex.

Reverse roles for second waltz step. Woman begins by turning onto her left foot. Man steps forward and between woman's feet.

It is possible to turn CCW with waltz steps, but this variation is not used here.

Jeg gikk meg over sjø og land

I Traveled over Sea and Land

Formation:	Boys and girls form a circle with low handhold. This dance is often done around a Christmas tree.
Steps:	Walking steps and other actions
	Note: Small children do not need to be concerned with specific footwork.

MUSIC 2/4

Phrase	Ms	Part
		I. CIRCLE CW
A	1-8	Beginning left foot, take 16 walking steps in CW LOD. On last step, pivot ½ CW turn on right foot.

Jeg	gikk meg	o-ver...
	1	2
	L	R

Phrase	Ms	Part
		II. CIRCLE CW
B	1-8	Children will now travel in CCW LOD and perform what they sing. Each verse features a different action. Everyone stamps during first verse:

Jeg hører	hjemm' i	tram-pe-	land, i...
1	2	1	2
stamp	stamp	stamp	stamp

Other actions are: pointing with finger, limping, hopping, and clapping.

REPEAT DANCE FROM BEGINNING

Klappdans

Clap Dance

Formation:	Several couples in circle B, facing CCW LOD. Join both hands (left hands together, right hands together) such that boy's right arm crosses behind girl's left arm.
Steps:	Change step
	Note: This is a simplified version of *clap dance.* Instructions for the usual version may be found in Klara Semb's book *Norske Folkedansar,* vol. 2.

MUSIC 2/4

Phrase	Ms	Part
		I. CHANGE STEPS AROUND CIRCLE; FORWARD AND BACK
A	1-7	Beginning left foot, dance 7 change steps forward in CCW LOD.
	8	Without releasing handhold, partners turn in place (boy CW, girl CCW) to face opposite direction with 3 steps, right-left-right. Partners may stamp with steps.
	9-15	Beginning left foot, dance 7 change steps forward in CW LOD.
	16	Release handhold. Partners turn in place (boy CCW, girl CW) to face each other with 3 steps, right-left-right. Partners may stamp with steps.
		II. BOW AND CURTSEY; CLAPS; TURN Boy folds arms above chest and parallel to ground. Girl holds skirt to either side.
B	1-2	Boy bows forward on ct 1 and comes up on ct 2. He then claps hands together 3 times on cts 1-&-2.
		Girl places ball of left foot behind right foot and curtsies on ct 1. She rises and brings her left foot back to place on ct 2. Girl then claps hands together 3 times on cts 1-&-2.
	3-4	Repeat action of ms 1-2, part II.
	5-6	Place left hand on hip. Partners scold each other 2 times with right pointing finger on cts 1-2. Place right hand on hip. Partners scold each other 2 times with left finger on cts 1-2.
	7-8	Clap hands together (as if hitting cymbals) on ct 1 and make 1 CCW turn in place with 4 steps.
	9-16	Repeat action of ms 1-8, part II.

REPEAT DANCE FROM BEGINNING

VARIATION

Older children may wish to vary the dance by substituting the following figure for ms 5-8, part II.

Ms	
5-6	Place left hand on hip. Partners clap right hands across on ct 1 and clap own hands together on ct 2. Place right hand on hip. Partners clap left hands across on ct 1 and own hands together on ct 2.
7-8	Slap partner's right hand across on ct 1 and make 1 CCW turn in place with 4 steps.

Sju gjenter går i ringen

Seven Girls Walk in the Ring

Formation: Equal number of boys and girls in 2 concentric circles. Girls form inner circle with low handhold and face center of circle. Boys form outer circle with low handhold and face center of circle.

Steps: Extending step, walking step

MUSIC 2/4

Phrase Ms Part

I. TWO CIRCLES

A 1-4 Beginning left foot, boys walk 8 steps in CW LOD.

Beginning left foot, girls walk 8 steps in CCW LOD.

Sju	gjen-ter	går i	rin-	gen . . .
	L	R	L	R

II. FIND PARTNER

5-8 Girls release handhold. Beginning left foot, they turn around and find a partner, walking 8 steps toward boys. Partners face each other and place both hands on hips.

Boys release handhold. They stand in place as girls come toward them. There should be some space between couples.

	det er	nett-opp	tin-	gen å
girls:	L	R	L	R
	fin-ne	ve-nen	sin.	Hei
	L	R	L	R

III. EXTENDING STEP AND TURNS

B 1-4 Stepping onto left foot first, perform 8 extending steps in place while facing partner.

hopp-sa-	sa fal-le-	ral-la-	-la!	Hei
L	R	L	R	
(extend R)	(extend L)	etc . . .		
hopp-sa-	sa fal-le-	ral-la-	-la!	For
L	R	L	R	

5-8 Clap hands together on ct 1 and then link right elbows with partner. Place left hand on hip, fingers forward. Beginning left foot, partners make 1 CW turn with 8 walking steps.

det er	nett-opp . . .
L	R

9-12 Release elbow hold. Repeat action of ms 1-4, part III.

13-16 Clap hands together on ct 1 and then link left elbows with partner. Place right hand on hip, fingers forward. Beginning left foot, partners make 1 CCW turn with 8 walking steps.

det er nett-opp . . .
L R

REPEAT DANCE FROM BEGINNING

Boys and girls may now trade places, such that boys form inner circle and girls outer circle. Boys will then choose partners during part II.

VARIATION

I composed this variation of *Sju gjenter* for children who were reluctant to choose partners.

New Formation: Equal number of boys and girls in 2 concentric circles. Girls form inner circle with low handhold and face **away** from center of circle. Boys form outer circle with low handhold and face **toward** center of circle. Each girl stands across from a boy and vice versa. They will be partners later in dance.

I. TWO CIRCLES

A 1-4 Beginning left foot, boys (or children in outer ring) walk 8 steps in CW LOD. On last step, pivot ½ CW turn on right foot.

Beginning left foot, girls (or children in inner ring) walk 8 steps in CCW LOD. On last step, pivot ½ CW turn on right foot.

5-8 Beginning left foot, boys walk 8 steps in CCW LOD.

Beginning left foot, girls walk 8 steps in CW LOD.

II. EXTENDING STEP AND TURNS

B 1-16 Place both hands on hips. Repeat action of ms. 1-16, part III.

Å jul med din glede

O Christmas With Joy

Formation: Boys and girls form a circle with low handhold. This dance is traditionally done around a Christmas tree.

Steps: Walking steps

MUSIC 4/4

Phrase	Ms	Part	
		I.	**CIRCLE CW AND CCW**
A	1-4		Beginning left foot, take 8 walking steps in CW LOD. On last step, pivot ½ CW turn on right foot.

Å	jul	med din	gle-	de og...
	1	2	3	4
	L		R	

Phrase	Ms	Part	
	5-8		Beginning left foot, take 8 walking steps in CCW LOD.

Vi	hil-	ser deg	al-	le med...
	1	2	3	4
	L		R	

Phrase	Ms	Part	
		II.	**CLAP, TURN CW, CURTSEY, BOW**
			Children will now face Christmas tree and perform what they sing.
B	1-4		Clap hands together 16 times on each beat of music.

Vi	klap-per i	hen-de-	ne,	vi...
	1	2	3	4
	clap	clap	clap	clap

Phrase	Ms	Part	
	5		Beginning right foot, make 1 CW turn in place with 4 steps.

Vi	sving-er	oss i	kret-	sen og
	1	2	3	4
	R	L	R	L

Phrase	Ms	Part	
	6		Everyone (including boys) places right foot behind left and performs a curtsey.

	nei- - - - - - -		er	og
	1	2	3	4
	R behind		R comes back to place	

Phrase	Ms	Part	
	7		Everyone (including girls) folds arms above chest and performs a bow.

	buk- - - - - - -		-ker.	
	1	2	3	4
bow	forward...		rise...	

REPEAT DANCE FROM BEGINNING

When you come to the third verse, the first part is performed differently. Everyone moves in a little closer to the Christmas tree. Cross your arms, such that you can give the person on your left your right hand, and the person on your right your left hand. Stand in place and move arms back and forth with a sawing motion. Part II is performed as before.

Song-Dances

Some Norwegian dances require using the voice as much as the feet. These are *songdanser* or "song-dances." Unlike other Norwegian dances, which are usually accompanied by fiddles or other instruments, the song-dances are accompanied by the singing of those who take part. This singing does not take a second place to the dancing; the songs and their contents are as much a part of the experience as the formations and steps.

Ballads and Steps

Many of the songs tell a story. This story is closely followed by the dancers and may involve a dozen or more verses, all of which are usually sung. Parts of the story are sometimes expressed through the way a song-dance is performed. A statement of defiance, for instance, may be emphasized with stamps.[1] A conversation may be divided into separate singing roles for the men and women.[2] Certain events may be represented with a gesture, as in the song-dance about Bendik and Årolilja. The dancers form what appears to be a blossoming flower by sweeping their arms upward. This symbolizes the lilies that spring from the graves of the two ill-fated lovers. Even when such movements are not used, the stories bring a sense of drama to the song-dances that is usually absent in other kinds of folk dances.

The songs cover a variety of both real and imaginary events from the near and distant past. Some are antiquated, such as "Terningspelet," which was probably first sung in Iceland during the 17th century.[3] It tells of a young man who won the hand of a princess through a game of dice. Closer to our own time is "Per Spelemann." Almost certainly from the 19th century, this popular song is about Per Spelemann or "Pete Fiddler," who claims that even if he becomes "as old as the stone under the bridge," he will never trade away his fiddle—not even for a cow.

The Norwegians sing their song-dances with gusto. Everyone is encouraged to learn the verses and sing in a wholehearted manner. Klara Semb, one of the key figures in the Norwegian song-dance movement, wrote plainly about this in her book, *Norske Folkedansar II:*

> *If you are to have spirit and life in the song-dances, it must be emphasized that all participants should know the verses and sing the entire time. Without this, one cannot enjoy the dances; if there are too many silent persons who do not bother to learn the verses or the melody, even those who know the songs will find no*

enjoyment in the dances. There will be too much dead weight to pull along.[4]

Most song-dances begin with a circle formation in which a group of couples knit themselves tightly together, like so many links in a chain, with one of several handholds. In such a formation there can be a tremendous feeling of momentum. When fifty or so dancers decide to move forward, one is obliged to go along for the ride.

Several step patterns may be used in the circle formation. Two of these, the *attersteg* and *kvilesteg,* are by far the most common. Neither of these names can be easily translated into English. Probably because he realized this, Johan Krogsæter created new names for his booklet, *Folk Dancing in Norway.* He called the *attersteg* the "open ballad step" and the *kvilesteg* the "closed ballad step," reflecting the relatively open and closed positions of the dancers when performing the two steps.[5]

The open ballad step is performed while facing the center of the circle with hands lifted to about shoulder level. Since the arms are lifted to either side of the body, with some space between one dancer and the next, this might be described as an open position. The footwork involves a pattern of four steps that move in a clockwise direction.

Open ballad step

The closed ballad step is done such that the dancers are angled toward the clockwise line of direction, almost facing the back of the person in front of them. They join hands at about waist level with an interlocking handhold. The arms are tucked in close to the body, with only a little space between one dancer and the next, making for a closed position. Like the open ballad step, the footwork of the closed ballad step involves a pattern of four steps that move in a clockwise direction.

The circle formation is often followed by a *brigde*. The word *brigde* means "change" and signifies a change to a new figure. This may be a chain (or "grand-right-and-left" as it is called in American square dance), steps in or out of the circle, a waltz, or other figures. The *brigde* usually occurs during the refrain of a given song. Once the *brigde* is complete, everyone repeats the song-dance from the beginning.

Various kinds of song-dances were once popular throughout Europe during the later Middle Ages, roughly from the mid-12th century to the mid-15th century. In England and France, for instance, there were song-dances called *caroles* that could be secular or religious in nature. An illustration of a *carole,* found in an early 14th century manuscript from France, shows a weaving line of dancers in which everyone has joined hands at about shoulder level.[6] Little else is known about the performance of *caroles,* although some have speculated that French *branles*—simple circle dances that were common during the Renaissance—may be related to *caroles*.[7]

A carole, accompanied by drum and bagpipes. From a 14th century French manuscript.
British Library

It is not easy to tell whether song-dances thrived in medieval Scandinavia. Accounts of dancing from that period are few and brief. Most are found in two collections of historical literature from Iceland, the Bishop Sagas and the Sturlunga Saga. These sagas refer to dances as a common amusement at weddings and other social gatherings during the 12th and 13th centuries. No descriptions were given, but a few passages speak of verses that were evidently chanted or sung when performing the dances. One example is mentioned in connection with the appointment of Bishop Jon to Northern Iceland in 1107:

> *This play was dear to men before the holy Jon was bishop, wherein a man would recite to a woman in the dance effeminate and satirical poems, and a woman to a man love verses.*[8]

There would appear to be only one account that makes a specific reference to song-dances in medieval Scandinavia. This is found in an early chronicle of church affairs in Denmark, which includes the following information about events in the Eskilsø cloister around the year 1175:

> *On yearly church holidays, numerous lay persons, including women, came to visit the monks at the cloister in order to celebrate the holidays with them. The lay persons were related to the monks. In the refectory they held a celebration, men and women together; they got drunk and did round-dances, accompanying themselves with song.*[9]

Because song-dances were once common in central Europe and probably in Iceland during medieval times, it seems entirely possible that they were once common in Norway as well. There is, however, no evidence of this today. Although several Norwegian authors have suggested that song-dances came to Norway from England around the year 1200, this is pure speculation. Scholars have been unable to find any historical accounts or surviving examples of song-dances in Norway. There are some examples of Norwegian *songleikar,* or "song-games," such as those that children perform around the Christmas tree, but it is hard to determine whether any of these might have had a connection with song-dances in Norway's past.

After the middle of the 15th century, song-dances went out of fashion in most of Europe. They were replaced by newer dances, especially by couple dances, which were typically accompanied by one or more musical instruments. Yet song-dances did not disappear altogether. They were reported at various times in some of the more remote parts of Europe, such as the Shetland Islands and Faeroe Islands.

It may be possible to catch a glimpse of what song-dances were like in Norway by looking at song-dances on the Shetland Islands and Faeroe Islands. Both groups of islands once had strong ties with Norway, and their cultural traditions were similar. Song-dances on the Faeroe Islands are of special interest. These led to the development of the song-dances that are done in Norway today.

The Shetland Islands lie to the north of Scotland and were settled by Norwegians during the 9th century. The islands flourished as a

Norwegian colony and outpost until Norway's properties went to the Danish crown in 1380. About a hundred years later, in 1472, the Danes ceded the islands to Scotland. Despite this alliance with Scotland, everyday customs were slow to change among the Shetland people. They continued, for instance, to speak Norn—a regional dialect of Old Norse—down through the 18th century.

Two sources mention song-dances in the Shetland Islands as late as the 18th and 19th centuries. Probably another remnant of Nordic culture, these song-dances were accompanied by a Norn *visick* (also spelled *viseck*). This appears to have been a ballad, according to the observations of George Low. In his travelogue of 1774, Low gave the first description of Shetland song-dances, noting that these differed from Scottish dances:

> *Diversions obtain only in winter, and consist in dancing on some stated days before and after Christmas . . . There is one species of dance which seems peculiar to themselves, in which they do not proceed from one end of the floor to another in a figure, nor is it after the manner of a reel; but a dozen or so form themselves into a circle, and taking each other by the hand, perform a sort of circular dance, one of the company all the while singing a Norn visick. This was formerly their only dance, but has now given entire way to the reel.*[10]

In 1882 Samuel Hibbert made a similar observation in his travelogue:

> *When the gue, an ancient two-stringed violin of the country, was aiding the conviviality of Yule, then would a number of happy sons and daughters of Yaltland take each other by their hands, and when one of them sang a Norn viseck, they would perform a circular dance, their steps continually changing with the tune.*[11]

Nothing remains of the Shetland song-dances today. They apparently faded from use sometime after Hibbert's visit in the early 19th century.

The Faeroe Islands are located about 375 miles off the southwest coast of Norway. Settled by Norwegians during the 9th century, they became a Norwegian province in 1035, but were transferred to Denmark in 1380.

In 1616, Jon Olafsson wrote of a song-dance he saw in the Faeroese town of Torshavn, which he described as "dancing in the round in the Faeroese manner with songs and ballads."[12] Thomas Jacobsen Tarnovious provided a more detailed report in a description of the Faeroe Islands from 1669:

> *The Faeroese exercise their bodies with various dances, where one holds the hand of the other, such that at one time there can well be twelve to fourteen persons out in the ring, with one part*

left open; then one of them sings the verse and they dance in response to the verses—sometimes fast, sometimes slow, sometimes with a stamp on the floor, sometimes with leg lifted high.[13]

Not long after Tarnovious wrote the above, Lucas Debes touched upon Faeroese song-dances in a work from 1673, this time identifying two occasions at which the dancing might take place:

...at their weddings and during the days of Christmas they entertained themselves with some simple dances, out in a circle, taking each other by the hand, and singing some old heroic ballads.[14]

Faeroese song-dance.
Norsk-Folkemuseum

Jens Kristian Svabo wrote about Faeroese dancing during his travels to the islands in the years 1781 and 1782. He confirmed the earlier accounts of a ring dance, but remarked that the ring could also be bent into a convoluted shape to accommodate more dancers.[15]

Faeroese song-dances weathered the centuries and are still done today. Hjalmar Thuren, a Danish ethnomusicologist, made a study of this tradition at the turn of the century. As those who wrote before him had discovered, Thuren found that Faeroese song-dances were usually performed in a circular or convoluted shape. The participants moved clockwise as they danced in a circle. They were led by a person called the *forsanger.* He led the singing as well, interpreting the verses by using vocal stress and perhaps some expressive gestures or more spirited movements. If the *forsanger* forgot a verse, it was possible to repeat the preceding verse, sometimes with an impromptu variation.[16]

Some *forsangers* had their own special songs, often passed down as family heirlooms, that would not be sung more than once a year. These songs might easily have more than 25 verses. Thuren noted that most of the song material consisted of antiquated ballads. He found that many of these ballads appeared to be medieval in origin, leading him to conclude the dances must have been of a similar age.[17] Thuren published his findings in two books: *Dans og Kvaddigtning paa Faerøerne* (Dance and Epic Poetry on the Faeroe Islands, 1901) and *Folkesangen paa Faerøerne* (Folk Songs on the Faeroe Islands, 1908). More recent research was presented by Astrid Luihn in her work, *Føryskur dansur* (Faeroese Dances, 1980).

Revived from the Middle Ages

The idea of doing song-dances in Norway originated with Hulda Garborg (1862-1934). Sometime during the early 1890s, Garborg attended a recital of traditional Norwegian folk songs by Thorvald Lammers. She was quite moved by this recital, writing later: "After I heard him sing those folk songs, I could never forget them; they sang their way into my soul for all time."[18] Garborg was dismayed that such songs had been all but forgotten in Norway and hoped to revive them in some way. It occurred to her that song-dances, such as those that were still done in the Faeroe Islands, might provide a means of doing this. "In the 1890s," she wrote in one of her books, "I began to think that the best way to bring the folk songs into use would be to bring the old song-dances into use."[20]

Hulda Garborg (1862-1934). Universitetsbiblioteket, Oslo.

Garborg soon discovered that very little had been written about Faeroese song-dances. It was not until January 1902 that she was able to find a rudimentary description of these, as given in Hjalmar Thuren's booklet, *Dans og Kvaddigtning paa Faerøerne.* With this description as her guide, she attempted to piece together some song-dances, using Norwegian songs for her accompaniment instead of Faeroese.[20-21]

Garborg taught her song-dances to a small group of women in Kristiania (now Oslo). Probably curious to see how the public would respond to this combination of traditional folk songs and circle dances, she arranged for her group to give a performance in Kristiania on the fourth of March, 1902. Her husband, the nationally-known author Arne Garborg, would assist at the performance by giving a talk about Nordic song-dances. His knowledge of the subject was probably based on the information in Thuren's booklet.

A sizeable crowd gathered to see the song-dances and hear Arne Garborg. Because there was not enough room for everyone at the place chosen for the performance, the Garborgs decided to postpone the event until the seventh of March, when they could use a larger hall. Several hundred people attended this later performance. Apparently quite impressed with the song-dances, the audience responded with much applause and enthusiasm.[22]

Several months after the performance in Kristiania, Garborg traveled to the Faeroe Islands. There she was able to gather more information about traditional song-dances and experience these in person. She returned to Norway in early July.[23]

Garborg took what she learned on the Faeroe Islands and used this as the foundation for a new generation of song-dances in Norway. These song-dances resembled Faeroese song-dances in many ways, but there were also some noticeable changes.

Faeroese song-dances were typically accompanied by ballads of considerable age. Norwegian song-dances, in contrast, were accompanied by a variety of songs, some of which were quite recent in origin.

Faeroese song-dances were rather simple and repetitive in nature. They had a certain primitive strength, especially when a number of experienced dancers were all moving and singing in unison, but there was little variety. Garborg wanted to preserve the original form of Faeroese song-dances in Norway. At the same time, she was not adverse to making some improvements. She wrote, "It is with dancing as with everything: new times call for new forms."[24] Garborg made the song-dances more interesting by adding new steps and figures. Such figures became known as the *brigde* section.

From the outset, Garborg had something more in mind that simply bringing Faeroese song-dances to Norway. Her aim was

A processional song-dance, Falkvor Lommansson, that was composed by Hulda Garborg and first performed in 1907.
Aftenposten photo archive.

clearly stated in the second edition of her book, *Songdansen i Norlandi,* published in 1913.

> *On the basis of song-dances from the Faeroe Islands, we have worked to bring into existence a new Norwegian song-dance that would be appropriate for our songs and our times.*[25]

During the next few years, Garborg visited different parts of Norway and taught song-dances. These were well-received and quickly grew in popularity. News of the song-dances even reached Sweden, where as early as 1903 Garborg was invited to give a presentation of song-dances at the Skansen open-air museum in Stockholm. She called upon a group of couples from Kristiania to help her with this presentation. Besides performing in Sweden, Garborg also taught song-dances there, using songs that had been collected and published in that country. Looking back on this experience, Garborg wrote:

> *We traveled [to Sweden] with ten couples, and people over there were so enthusiastic about the song-dances that in the following year they created their own dance group with Swedish youth, whom I have had the great pleasure to work with. I found appropriate Swedish dance songs from Afzelius and Arvidson. The situation in Sweden was the same as here: everyone had forgotten their old songs. Both young and old were happy to learn these on the "second time around." And let me say, there was singing then!*[25]

The Swedes were so taken with the song-dances that they sponsored what might be called the first international conference for song-dances. This was held at a school in Naas in 1905 and included participants from eleven countries.

The great success of the song-dances is not surprising when viewed in light of current trends. The 19th century was a time of rapid change in Scandinavia. Many of the old folk traditions were being left behind in the wake of the industrial revolution. Around the turn of the 20th century, people began to take a special interest in these traditions and sought ways to preserve them, from organizing handicraft societies to building folk museums. Garborg was one of the first pioneers in this effort. Besides her work with folk songs, she researched Norwegian folk costumes. Such costumes were rarely worn in the rural districts and were almost unknown to people in the towns of her time. Her research resulted in an illustrated book, *Norsk Klædebunad,* published in 1903. It was the first resource of its kind in Norway.

The song-dances also coincided with the introduction of a new and pervasive style of art in Scandinavia. This was the *Jugendstil.* Corresponding to what other Europeans called the *art nouveau* style, the *Jugendstil* was influenced by works from different historical periods, but those from the Middle Ages were among the most important.[26]

Gerhard Munthe was the foremost representative of the *Jugendstil* in Norway. Although Munthe was mostly concerned with visual art and interior design, his works would have something important in common with Garborg's song-dances. Both were inspired by medieval designs and Norwegian folk culture in general. For instance, while Munthe painted several scenes from the medieval ballad "Åsmund Frægdegjæva"between 1902 and 1904,[27] Garborg composed a dance to it, probably in 1911 or 1912.[28]

An Expression of Nationalism

Another significant trend was Norway's movement for national independence. Although the country had become separate from Denmark after the Napoleonic wars in 1814, ending nearly four centuries of foreign rule, it was then given to Sweden as a self-governing territory. The Norwegians were not able to break their ties with Sweden until 1905.

The movement for national independence served to remind the Norwegians of their cultural heritage. Reviving songs from Norway's past by means of the song-dances became a way of proclaiming that heritage. It underscored the traditional life and language of Norwegians at a time when they were seeking national themes to rally around.

It should be mentioned that language was a significant issue during the movement for national independence. During the years of Danish rule, the Danish language had become the written language of Norway. Educated people, particularly those in the cities, spoke a blend of Danish and Norwegian. At the turn of the century, there was a widespread campaign to promote a more pure form of the Norwegian language in both writing and speaking. The song-dances furthered the aims of this campaign by presenting songs that used only Norwegian words.

More than the invention of a single woman, Norwegian song-dances were a product of their time. They reflected the events of a particular period in Norwegian history, namely, the years between 1890 and 1910. From our perspective today, we may accurately describe them as historical dances, even though they are not especially old.

Garborg had many opportunities to promote song-dances while on tour in Norway with *Det Norske Spellaget* (the Norwegian Acting Company). She was the director of this company in 1911 and 1912. Each play was followed by a display of song-dances, performed by four couples from the company dressed in national costumes.[29]

Garborg wrote three works that dealt with song-dances. Two of these were *Norske Folkevisor* (Norwegian Folk Songs) and *Norske Dansevisur* (Norwegian Dance Songs). First published in 1903 and 1913, respectively, the two featured various Norwegian folk songs with some instructions for song-dances. The third work was *Song-dansen i Nordlandi* (The Song-dance in the North Lands), which was published in three editions in 1903, 1913, and 1914. Part of this booklet related Garborg's early experiences with the song-dances. The remainder gave a brief discussion about dance and song-dances in other countries, from Greece to the Faeroe Islands. By the time the second edition of the booklet was published in 1913, she was able to include a favorable report of her progress with Norwegian song-dances and folk songs:

> *And I believe we have come a long way. The dance is not firmly established in all places yet, but there are several groups of youth around the country who are already good at it, and the Norwegian songs have become more and more known.*[30]

Garborg's involvement with the song-dances appears to have declined after 1913. The movement was carried on by Klara Semb (1884-1970). In addition to being one of the dancers who accompanied Garborg on the Stockholm tour in 1903, Semb was one of the actors who performed song-dances with *Det Norske Spellaget*. Already in 1903, Semb was giving song-dance workshops in Romerike, a region immediately outside Kristiania. These workshops soon led to others. Like Garborg, Semb began composing her own song-dances, at first using medieval ballads for her accompaniment. She traveled to the Faeroe Islands in 1911 in order to learn more about song-dances there.[31]

Semb was also interested in folk dances that were still being done in the Norwegian countryside. Starting in 1915, she traveled to different parts of the country and began to learn and record these dances. Most were group dances, such as reels, which involved several people in various changing formations. Semb called these *turdanser,* or "figure-dances."

Several hundred Norwegians perform a song-dance in concentric circles at a gathering in Sunnfjord.

Semb decided to present a collection of dances in a series of books called *Norske Folkedansar* (Norwegian Folk Dances). The first volume (1922) contained a collection of songs for the song-dances, the second (1922) contained instructions for the song-dances and figure-dances, and the third (1924) contained music for the figure-dances. In later editions, Semb added some dances and discarded others. She continued to compose song-dances, drawing upon her knowledge of Faeroese and Norwegian dances. Over 100 song-dances were published in the fifth edition of *Norske folkedansar II* in 1956. Of these, nearly 70 percent were her compositions. She also created song-dances for children and presented these in her book, *Danse, danse, dokka mi* (1958).

Semb's work was not confined to collecting, notating, and composing dances. She also taught these in workshops and led a folk dance group in Oslo. Her efforts quickly came to the attention of Noregs Ungdomslag, or "Norway's Youth League." This league, founded in 1896, sought to involve young people in healthful activities, especially those that stressed Norway's cultural heritage. In 1921, Noregs Ungdomslag formed a council to promote folk dancing among its membership. For example, in 1922 the organization sponsored a teachers' workshop in folk dancing. Semb was asked to lead the workshop.[32]

This was only the beginning of a long relationship between Semb and Noregs Ungdomslag. More workshops followed and the dances spread widely. The organization continued to support Semb's work until she retired at age 70 in 1954. At this time the Norwegian parliament made Semb a Knight of St. Olav's Order for her contributions to folk dance and folk music.

Much as Garborg had hoped, the song-dances have proved to be an excellent way of keeping traditional songs alive in Norway. They have given the average person an opportunity to learn and enjoy these songs. But more than this, the song-dances have become a popular dance tradition in their own right, now coexisting with older traditions. Hundreds of Norwegians may be seen performing song-dances at various dance events and folk festivals. Most importantly, new generations of young people have stepped in to carry the tradition further. Some Norwegians have followed the example of Garborg and Semb by adding to the repertoire with song-dances of their own.

Notes

1 As in the song-dance, *Eg gjekk meg opp til sæterli.*

2 As in the song-dances, *Kjære du mi Randi* and *Trolldans.*

3 Klara Semb, *Norske Folkedansar,* vol. II, 5th ed., (Oslo: Noregs Boklag, 1956; reprint 1975), p. 151.

4 Ibid., p. 9.

5 Johan Krogsæter, *Folk Dancing in Norway* (Oslo: Johan Grundt Tanum Forlag, 1968), p. 16.

6 Mary Clarke and Clement Crisp, *The History of Dance* (New York: Crown Publishers, 1981), p. 120.

7 *The New Grove Dictionary of Music and Musicians,* 1980 ed., s.v. "Branle," by Daniel Heartz.

8 Gudbrand Vigusson and York Powel, ed. and trans., "Ioan's Saga," *Origines Islandicae* (Oxford: Clarendon Press, 1905), vol. I, pp. 554-55.

9 H. Grüner Nielsen, "Dans i Danmark," *Nordisk Kultur,* vol. 24 (Stockholm: Albert Bonniers Forlag, 1933), pp. 132-33.

10 George Low, *A Tour Through Orkney and Shetland* (n.p., 1774; reprint ed., Kirkwall: William Peace & Son, 1879), p. 163.

11 Samuel Hibbert, *A Description of the Shetland Islands* (Edinburgh: 1822) cited by Otto Andersson, "The Bowed Harp," *Galpin Society Journal* 23 (1970): 20-21.

12 H. Grüner Nielsen, "Dans Paa Færøerne," *Nordisk Kultur,* vol. 24 (Stockholm: Albert Bonniers Forlag, 1933), p. 153.

13 Thomas J. Tarnovius, *Ferøers Beskrivelser udi største korthed* (n.p.: NKS, 1949), p. 75.

14 Lucas Debes, *Færoa reserata* (Copenhagen: 1673), p. 251.

15 Jens Kristian Svabo, *Reise i Færøe i Aarene 1781 og 1782,* cited by Hjalmer Thuren, *Folkesangen Paa Færøerne* (Copenhagen: Andr., Fred., Høst., & Sønner Forlag, 1908), p. 17.

16 Thuren, *Folkesangen,* pp. 8-25, 31.

17 Ibid., pp. 35-44.

18 Garborg, *Songdansen i Nordlandi* (Kristiania: H. Aschehoug & Co., 1913), p. 2.

19 Ibid., p. 9.

20 Ibid., pp. 9-10.

21 Egil Bakka, "Songdansen," in *Norske Folkedansar: Songdansar,* rev. ed., by Klara Semb et al., (Oslo: Noregs Boklag/Det Norske Samlaget, 1985), p. 21.

22 Garborg, *Songdansen,* p. 10.

23 Ibid., p. 11.

24 Ibid., p. 10.

25 Ibid., p. 12.

26 H. H. Arnason, *History of Modern Art* (Englewood Cliffs: Prentice-Hall, Inc., 1977), pp. 72-73.

27 Henning Alsvik and Leif Østby, *Norges Billedkunst i det nittende og tyvende århundre* (Oslo: Gyldendal Norsk Forlag, 1951), pp. 328-332.

28 Semb et al., *Norske Folkedansar: Songdansar,* p. 179.

29 *Norsk Allkunnebok,* 1953 ed., s.v. "Garborg, Hulda." *Gyldendals store konversasjonsleksikon,* 3rd ed., (1972), s.v. "Garborg, Hulda," by Sigmund Moren. Liv Midttun, *Klara Semb: Biografiske data og fakta* (Oslo: Noregs Boklag, 1981), p.4.

30 Garborg, *Songdansen,* p. 10.

31 Liv Midttun, *Biografiske data og fakta om Klara Semb* (Oslo: Noregs Boklag, 1981), pp. 3-4.

32 Ibid., p. 5.

Den bakvende visa

The Backwards Song

Formation: Several couples in circle A, facing center circle with open handhold.

Steps: Light step, open ballad step , step-hop

MUSIC 6/8 ______________________________

Phrase	Ms	Part	
		I.	**OPEN BALLAD**
A	1-8		Perform 2 open ballad patterns and first 3 steps of another open ballad pattern. Following this, close right foot beside left and place weight on right foot.

Eg	beis- la min	støv-el og	sa- la mit	sverd,	så	batt eg	mer- ra ved
	L	R	L	l		R	r
	si -	da. så	reis- te eg	meg	til	kat-te-mords-ferd	skulle
	L	R	L	l		R	r
	stan-de den	far- leg- e	stri -	da.	Eg		
	L	R	L	R close			

II. MAN BENDS AND TURNS CW

Release handhold with corner person. Man changes handhold with partner, lightly clasping fingers of her left hand from above with his right. He lowers left arm by his side. Woman lowers right hand and holds skirt to side.

9 Man bends at waist and goes under woman's outstretched arm. Beginning left foot, he makes 1 CW turn with 3 steps. As he begins to turn, he puts his right arm behind him, such that he presses back of right hand against his lower back. He continues to hold woman's left hand during turn.

Beginning left foot, woman performs 3 steps in place. At same time, she extends left arm, allowing man to turn beneath it.

låg	og	eg	datt,		eg
1	2	3	4	5	6
L		R	L		

Note: This part is performed more slowly than part I. Sing words such that man has enough time to turn comfortably beneath arm.

III. WOMAN TURNS CW

10 Man straightens up and faces partner. Beginning right foot, he performs 3 steps in place. At same time, he lifts right hand, such that woman can turn beneath it.

Beginning right foot, woman makes about 1 CW turn beneath uplifted arms with 3 steps.

drøym-	de	i	natt,		eg
1	2	3	4	5	6
R		L	R		

Note: Tempo is same as in part II.

IV. COUPLE TURNS CW

Partners assume position 4. Man is angled toward center of circle, woman away.

11-12 Beginning left foot, make 1 CW turn with 3 step-hops in a triple rhythm (see instructions for step-hop). Partners begin to open away from each other with last step-hop. Woman turns somewhat CW, such that she will be to right side of man in circle.

tot-	te	den	vi-	sa	var	bak-		vend
1	2	3	4	5	6	1	2	3
L		l	R		r	L		l

Taking 1 step on right foot, partners face center of circle. Arms are lowered as everyone comes into place.

radt______.		
4	5	6
R		

Note: Return to quicker tempo of part I. Partners should be able to do step-hops at a natural pace.

REPEAT DANCE FROM BEGINNING

Eg gjekk meg opp til sæter li

I Went Up to the Mountain-farm Meadow

Formation:	Several couples in circle A, facing center of circle with open handhold.
Steps:	Change step, light step, open ballad step
MUSIC 6/8	___

Phrase	Ms	Part	
		I.	**OPEN BALLAD**
A	1-12		Perform 4 open ballad patterns.

Eg gjekk meg	opp	til sæ-	ter- li	til jen-	ta mi,
L	R	L	l	R	r
ho var ik- kje	blid___				.
L	R	L	l	R	r

Repeat verse and footwork.

II. MEN TOWARD CENTER OF CIRCLE: WOMEN SCOLD

Release handhold. Man hooks thumbs in armholes of vest. Woman places left hand on hip, fingers forward.

B 1-2 Beginning left foot, man performs 2 change steps toward center of circle. On last step of second pattern, he pivots on right foot and makes about ½ CW turn. He concludes facing partner.

Woman stands in place. Looking at partner, she bends right arm and scolds him 2 times with right finger. Woman then repeats action with other hand, placing right hand on hip and scolding with left finger 2 times.

	Ho var	ik-	kje	grei,		ho
	1	2	3	4	5	6
man:	L		R	L		
woman:	scold			scold		

	sva-	ra	meg	nei,		kva
	1	2	3	4	5	6
man:	R		L	R (pivot)		
woman:	scold			scold		

Note: Rhythm for man's change step may be described as slow-quick-slow.

III. MAN DANCES TOWARD PARTNER; WOMAN TURNS CCW

3-4 Beginning left foot, man performs 1 change step toward partner, traveling away from center of circle. He then stamps 4 times in place in front of partner with right-left-right-left.

Beginning right foot, woman makes 1 CCW turn in place with 2 light steps. She claps hands together on first turning step. Woman then stamps 4 times in place with left-right-left-right. At same time, she places both hands on hips, fingers forward.

	vil	du	no	her,		du
	1	2	3	4	5	6
man:	L		R	L		
woman:	L			R		
	clap					

	fær	ik-	kje	meg____.		
	1	2	3	4	5	6
man:	R	L	R	L		
woman:	L	R	L	R		

IV. SCOLDING

5-6 Partners stand in place. They scold each other 2 times with right finger and then 2 times with left finger. Woman's other hand is placed on hip, fingers forward. Man's thumb of other hand is hooked in armhole of vest.

Ho	var	ik-	kje	grei,		ho
	1	2	3	4	5	6
	scold			scold		

sva-	ra	meg	nei,		kva
1	2	3	4	5	6
scold			scold		

V. MAN AND WOMAN TURN CCW

7-8 Beginning left foot, make 1 CCW turn in place with 2 light steps. Clap hands together on first turning step. Stamp 4 times in place with left-right-left-right. At same time, man puts thumbs in armholes of vest and woman places both hands on hips, fingers forward.

vil	du	no	her,		du
1	2	3	4	5	6
L			R		
clap					

fær	ik-	kje	meg____.		
1	2	3	4	5	6
L	R	L	R		

REPEAT DANCE FROM BEGINNING

To begin dance again, man makes ½ CW turn with first open ballad pattern, joining partner in circle.

Eg gjætte Tulla

I Watched Over Tulla

Formation: Several couples in circle A, facing CW LOD with closed handhold.

Steps: Closed ballad step, light step.

MUSIC 3/4 ______________________________

Phrase	Ms	Part

I. CLOSED BALLAD

A 1-8 Perform 4 closed ballad patterns.

Eg	gjæt-	te	Tul-	la	i	fem-	ten	år______	
&	1		2	3		1		2	3
	L		R	L		l		R	r
eg	pas-	sa	vel		på	Tul-		la______.	
&	1		2	3		1		2	3
	L		R	L		l		R	r
eg	gjæt-	te	bort	bå-de		lam	og	får______	
&	1		2	3		1		2	3
	L		R	L		l		R	r
men	en-	då	had-	de	eg	Tul-		la______.	
&	1		2	3		1		2	3
	L		R	L		l		R	r

II. SIDE STEPS WITH DIPPING ACTION

Release handhold with corner person and turn to face center of circle. Man holds partner's left hand from beneath with his right hand, such that her fingers rest on top of his. Man bends right arm and draws woman's left hand across to upper left side of his chest. Partners should have eye contact with each other.

9 Beginning left foot, take 1 step sideways to left, bending at ankle and knee. On next ct., close right foot beside left and transfer weight to both feet. Rise to balls of feet as right foot comes into place. With weight still on both feet, perform 2 dips in place, bending and extending ankles and knees.

Å	hei	ă	hå______.	
&	1		2	3
	L		B	B
			dip	dip

10 Repeat action of ms 9. part II.

det	fær	så	gå ______.	
&	1		2	3
	L		B	B
			dip	dip

III. COUPLE TURNS CW

Partners turn to face each other and join hands with position 6.

11-12 Beginning left foot, partners make 1 CW turn in place with 6 walking steps.

Eg	tre- gar	mes-	te på	Tul-	la______.	
&	1	2	3	1	2	3
	L	R	L	R	L	R

IV. WOMAN TURNS CW

Partners should be facing each other.

13-14 Man releases left hand from partner's right. Beginning left foot, man performs 6 small steps in place. At same time, he lifts right hand, such that woman can turn beneath it.

Beginning left foot, woman makes 2 CW turns beneath uplifted arms with 6 small steps in place.

for	ho var	krul-	la i	ul-	la______.	
&	1	2	3	1	2	3
	L	R	L	R	L	R

REPEAT DANCE FROM BEGINNING

To begin dance again, release handhold and face CW LOD (man makes ½ CCW turn in place). Assume closed handhold.

Eg rodde meg ut på seie grunnen

I Rowed Out to the Fishing Place

Formation: Several couples in circle A, facing CW LOD with closed handhold.

Steps: Closed ballad step, step-hop

MUSIC 3/4 ________________________________

Phrase Ms Part

I. CLOSED BALLAD

A 1-8 Perform 3 closed ballad patterns and first 3 steps of another closed ballad pattern. Following this, release closed handhold and face partner (man makes ½ CW turn, woman dances in place) with 2 steps, right-left.

Eg	rod-de meg	ut	på	sei-e	grun-	nen
	1	2	3	1	2	3
	L	R	L	l	R	r

det	var om	mor-	go-nen	tid-	leg________	.
	1	2	3	1	2	3
	L	R	L	l	R	r

Då	kom han	O-	lav frå	Kå-re-	mun-	nen
	1	2	3	1	2	3
	L	R	L	l	R	r

og	lag- de	bå-	ten for	i - - -
	1	2	3	1
	L	R	L	l

le________	.
2	3
R	L
step	step

II. SLAPPING

Stand in place.

B 1 Man swings right arm and pretends to slap woman on her left cheek. At same time, woman pretends as if she had been hit; she moves head to right, bends down somewhat to right, and claps her left hand to her right hand, which is down by her right side.

Note. Clapping action provides "sound effect" for slap; it should be done cleverly so that action is not obvious.

	Då	dreiv eg	til	han med
		1	2	3
man:		slap		

2 Woman swings right arm and pretends to slap man on his left cheek. At same time, man pretends as if he had been hit; he moves head to right, bends down somewhat to right, and claps his left hand to his right hand, which is down by his right side.

	fis- ke	ston-	gi, så'n
	1	2	3
woman:	slap		

3 Repeat action of ms 1, part II.

	datt i	u-	vit- et
	1	2	3
man:	slap		

4 Repeat action of ms 2, part II.

	bak i	ron-	gi______ .
	1	2	3
woman:	slap		

On last extended note, man makes ½ CCW turn in place. Everyone faces CCW LOD and assumes closed handhold. Place weight on right foot.

III. CLOSED BALLAD

5-8 Perform 2 closed ballad patterns.

Eg	vart så	glad	tok	til og	kvad	
&	1	2	3	1	2	3
	L	R	L	l	R	r
eg	råd-de	grun-	nen	å- lei-	ne______ .	
	1	2	3	1	2	3
	L	R	L	l	R	r

IV. COUPLE TURNS CW

Release handhold. Man turns a little CW, such that he is angled toward center of circle. Woman turns somewhat CCW, such that she is angled away from center of circle.

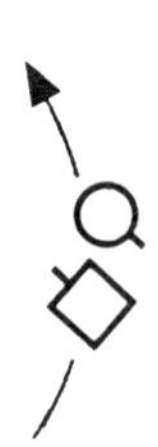

Assume following position:

9 Beginning left foot, make 1 CW turn with 3 step-hops in a triple rhythm (see instructions for step-hop). Partners begin to open away from each other with last step-hop. Woman turns somewhat CW, such that she will be to right side of man in circle.

3			3			3		
Su-	de-	li	Su-	de-	li	Su-	de-	li-
L		l	R		r	L		l

10 Taking 1 step on right foot, partners face center of circle and quickly lower arms. Everyone joins hands. On next ct, swing arms forward and raise hands above head level, standing on both feet.

-dei	hoi______ .
R	B

REPEAT DANCE FROM BEGINNING

To begin dance again, face CW LOD and assume closed handhold.

Fram dansar ein haugkall*

A Hill-man Dances Forth

Formation: Several couples in circle A, facing CW LOD with closed handhold.

Steps: Closed ballad step, reinlender step

*A haugkall is a mythical being who lives underground.

MUSIC 4/4 ______________________________

Phrase	Ms	Part
		I. CLOSED BALLAD
A	1-8	Perform 5 closed ballad patterns. Following this, release closed handhold and face partner (man makes ½ CW turn, woman dances in place) with 2 steps, left-right. Note that second step occurs on a note that leads into next part of song.

Fram	dan-	sar ein	haug-	kall	fa-	ger og
	L	R	L	l	R	r
	blå	med	gull-	ring om	hå-	ret som
	L	R	L	l	R	r

fløy_ _	-- - -	- mer.	Han	gil-	jar	for
L	R	L	l	R	r	
Ves-	le-møy	til	og	frå		og
L	R	L	l	R	r	
to-	nar	i-kring	hon-om	strøy-	- -	
L	R	L	l	R	r	
-mer	Å.					
L	R					
face partner						

II. CHAIN

B 1-3 Take right hands with partner on ct 1 and chain past 3 persons, counting partner as number 1. Beginning left foot, dance 1 reinlender step per person. Men move in CCW LOD, women in CW LOD.

4 Continue to chain to fourth person, giving him or her your left hand. At same time, woman makes ¾ CCW turn in place with 1 reinlender step, right-left-right. Woman's left arm will cross in front of her as she turns. Man comes to right side of fourth woman with 1 reinlender step, right-left-right. Everyone should conclude in circle formation.

hild-	ran-de	du		med
1	2	3	4	
L	R	L	l	
meg	skal du	bu		i
1	2	3	4	
R	L	R	r	
blå	haug-en	skal	du	din
1	2	3	4	
L	R	L	l	
sylv-	rokk	snu______.		
1	2	3		
R	L	R		

REPEAT DANCE FROM BEGINNING

To begin dance again, release left hands and face CW LOD with closed handhold.

Hei huskom i hei

Hey, What a Hullabaloo

Formation: Several couples in circle A, generally facing CW LOD. Man stands in front of his partner and turns upper body a little CW. He extends his right arm backward and holds his partner's left hand with his right. His left arm is by his side. Woman stands behind partner and somewhat to inside of circle. Her free arm is relaxed by her side.

Steps: Change step, open ballad step, walking step

MUSIC 2/4

Phrase Ms Part

I. STAMP

With right foot, take 1 step forward in CW LOD and stamp on that step. This occurs on introductory note before first ms.

Hei
R

II. WOMAN TURNS CW; WALKING STEPS

A 1 Beginning left foot, man performs 1 change step forward. At same time, he briefly turns a little toward center of circle and lifts right hand, such that woman can turn beneath it.

Beginning left foot, woman makes 1 CW turn beneath uplifted arms with 1 change step. Footwork travels forward.

husk- om	i	hei	sa
1	&	2	
L	R	L	

2 After turning action by woman, man lowers right hand to comfortable position by his right shoulder, arm bent. Woman continues to hold man's hand with her left, extending left arm forward. She faces CW LOD.

Beginning right foot, perform 2 walking steps forward with some spring in each step.

hal-	lin-	gen,	eg
1	&	2	
R		L	

III. MAN TURNS CCW; WALKING STEPS

3 Beginning right foot, woman performs 1 change step forward. At same time, she lifts left hand, such that man can turn beneath it.

Beginning right foot, man makes 1 CCW turn beneath uplifted arms with 1 change step. Footwork travels forward.

rug-	gar meg	sjøl
1	&	2
R	L	R

4 After his turn, man lowers right hand to comfortable position by his right shoulder, as before. Woman continues to hold man's hand with her left.

Beginning left foot, perform 2 walking steps forward with some spring in each step.

i kvel-	lin-	gen	med
1	&	2	
L		R	

IV. OPEN BALLAD

Turn to face center of circle and assume open handhold.

5-11 Perform 2 open ballad patterns. Following this, release open handhold and turn to face CW LOD. Partners join hands (man's right, woman's left) as at beginning of dance. Take one step forward on left foot.

ha-mar og	tong sa	mo-	in-gen Kok	vel-ling på	graut sa
L	R	L	l	R	r

sig-dø-ling-	gen så	sma-kar eg	med sa	kryl-lin-	gen_____
L	R	L	l	R	r

_____ .
L
step

REPEAT DANCE FROM BEGINNING

Ola Glomstulen

(man's name)

Formation: Several couples in circle A, facing center of circle. Dancers are close beside each other. They assume following handhold. Men extend arms to either side, reaching in front of women, and take hands. Women extend arms to either side, reaching in front of men, and take hands. Your left arm should be beneath arm of person to your left. Your right arm should be above arm of person to your right.

Steps: Change step, (¾ rhythm, second version) light running step, light step

MUSIC 3/4 ________________________________

Phrase	Ms	Part
		I. CIRCLE CW AND CCW
A	1-4	Beginning left foot, perform 4 change steps in CW LOD. Footwork travels forward in CW LOD, but upper body is turned somewhat toward center of circle. On last ct, pivot ½ CW turn on right foot without releasing handhold.

O-la	Glom-	stu-	len
	L	R	L
	had-	de	ei
	R	L	R
	ga-	mal	grå
	L	R	L
	geit_______.		O- la
	R	L	R
			pivot

Phrase	Ms	Part
	5-8	Beginning left foot, perform 4 change steps in CCW LOD. Footwork travels forward in CCW LOD, but upper body is turned somewhat toward center of circle. On last ct, pivot on right foot to face center of circle.

Glom-	stu-	len
L	R	L
had-	de	ei
R	L	R
ga-	mal	grå
L	R	L
geit_______.		Å
R	L	R
		pivot

II. EVERYONE FORWARD AND BACK

B 1-2 Beginning left foot, perform 2 small change steps toward center of circle. Hands should lower a little as everyone dances forward, allowing circle to become smaller.

Kjæ-	re	mi
L	R	L
Ka-	ri	gjer
R	L	R

3-4 Beginning left foot, perform 1 large change step away from center of circle. On first ct of next ms, close right foot beside left.

pøl-	sa	vel
L	R	L
feit______.		For i
R		
close		

III. COUPLE TURNS CCW; WOMAN TURNS CCW

Release handhold and assume following position:

5-6 Partners make 1 CCW turn in place as follows.

Man dances 1 step backward on left foot (ct 1) and then 1 step-hop backward on right foot (cts 2 - 3). His left foot should lift a little backward during step-hop. During next ms, man repeats same pattern, completing turn.

Beginning left foot, woman performs 6 light running steps forward, turning with partner as she does so.

	mor-	gon skal	O-la	Glom-	stu-	len
	1	2	3	1	2	3
man:	L	R	r	L	R	r
woman:	L	R	L	R	L	R

7-8 Beginning left foot, man performs 4 light steps in place. Without releasing handhold, he will, at same time, lift his left arm over his head and then his right arm over woman's head, such that he helps her to make 1 CCW turn.

Beginning left foot, woman makes 1 CCW turn under uplifted arms with 4 light steps.

gif- - - - - - - ta   seg____ .
1   2   3   1   2
L   R   L   R

Partners should conclude facing center of circle with following position:

REPEAT DANCE FROM BEGINNING

Terningspelet

The Dice Game

Formation: Several couples in circle A, facing center of circle with open handhold.

Steps: Light step, open ballad step, walking step

MUSIC 4/4 ______________________________

Phrase	Ms	Part
		I. OPEN BALLAD
A	1-4	Perform 2 open ballad patterns and first 3 steps of another open ballad pattern. Following this, close right foot beside left with weight.

Prin-   ses- sa   sat i   hø-gan   loft med   han- di   under
         L   R   L   l   R   r

kinn____.   Da   kom ein   fag- er   un- ger-   svein
L   R   L   l   R   r

gan- gan-de dit   inn____.
L   R   L   R
             (close)

II. SIDE STEPS; SWING LEG

5 Beginning left foot, perform 1 light step to your left and then cross right foot behind it with weight. Take another step to your left and then swing right foot in front of left, such that right leg extends naturally. Right foot should point generally toward floor.

For______	dei	lei-	ka
1	2	3	4
L	R	L	l
side	behind	side	swing r foot

6 Repeat action of ms 5 in opposite direction with opposite footwork.

for______	dei	lei-	ka, dei
1	2	3	4
R	L	R	r
side	behind	side	swing l foot

III. COUPLE TURNS CW; INFORMAL GREETING

See instructions for informal greeting.

7 Partners turn to face each other and join hands across in position 6. Beginning left foot, make ¾ CW turn with 4 walking steps. Man should conclude facing toward center of circle, woman away.

lei- ka	og dei	spe- - -	la gull-
1	2	3	4
L	R	L	R

8 Man places weight on both feet. With arms still extended in position 6, he bows toward partner in 2 cts. Allow arms and hands to follow motion of upper body in a natural manner. Man rises in about 2 cts (there is only one ct written in music, but this is extended, giving more time to rise from bow).

Swinging left foot behing right, woman lowers with informal curtsey in 2 cts. She rises in about 2 cts.

-ter- - - - - - - - -		ning____.
1	2	3
lower		rise

REPEAT DANCE FROM BEGINNING

To begin dance again, man releases left hand and dances a little forward into circle with first open-ballad pattern. Woman releases right hand and makes ½ CW turn with first open ballad pattern, joining partner in circle.

Trolldans

Troll Dance

Formation: Several couples in circle A, facing CW LOD with closed handhold.

Steps: Closed ballad step, reinlender step, step-hop

MUSIC 2/4 ______________________________

Phrase Ms Part

I. CLOSED BALLAD

A 1-6 Perform 4 closed ballad patterns.

Det	lei-	kar så	mangt	ei	su-	mar-
	L	R	L	l	R	r
	nott,	i	dver-	ge	skog____	i
	L	R	L	l	R	r
	dver-	ge	skog____.		Al-	ve-
	L	R	L	l	R	r
	dans	og	hul-	dre-	lått,	å
	L	R	L	l	R	r

II. WOMAN DANCES TOWARD CENTER OF CIRCLE

Loosen handhold, lightly grasping hands of person on either side of you.

7 Beginning left foot, man performs 2 step-hops, traveling only a little forward. At same time, he extends arms, allowing women on either side of him to travel toward center of circle. His upper body should be angled toward center of circle.

Beginning left foot, woman performs 1 reinlender step on a diagonal, such that she travels toward center of circle and leads with her left shoulder. She looks toward corner person. On last ct, she turns a little CCW.

	hei	deg å	hild-	rande
	1	&	2	&
man:	L	l	R	r
woman:	L	R	L	l
				turn

III. DANCING TOWARD/AWAY FROM CENTER OF CIRCLE

8-14 Perform 7 reinlender steps, such that everyone travels in a zig-zag pattern around circle in CW direction. Footwork is given below.

Beginning left foot, man performs first reinlender step on a diagonal, such that he travels toward center of circle and leads with his left shoulder. He passes by partner (who is now traveling **away** from center of circle) and acknowledges her. On last ct, he turns a little CCW.

Beginning right foot, man performs second reinlender step on a diagonal, such that he travels away from center of circle and leads with his left shoulder. He passes by corner person (who is now traveling **toward** center of circle) and acknowledges her. On last ct, he turns a little CW.

Man continues to dance toward and away from center of circle. On his seventh reinlender step, he dances mostly in place, traveling only a little forward in CCW LOD.

Woman dances contrariwise in relation to man. Beginning right foot, she performs first reinlender step away from center of circle. Her seventh change step travels away from center of circle. She should conclude next to partner in circle.

Note: Arms should be held in relaxed manner, such that they can extend as men and women travel apart from each other.

	du ______	.		
	1	&	2	&
man:	L	R	L	l
woman:	R	L	R	r
	Lett-	le- ge	trø-	da me
	1	&	2	&
	R	L	R	r
	L	R	L	l
	dan-	sen i	kveld	un-der
	1	&	2	&
	L	R	L	l
	R	L	R	r
	fjel - -	- - -	li________	.
	1	&	2	&
	R	L	R	r
	L	R	L	l
	Lett-	le-ge	trø-	da me
	1	&	2	&
	L	R	L	l
	R	L	R	r
	dan-	sen i	kveld	un-der
	1	&	2	&
	R	L	R	r
	L	R	L	l
	him- -	- - -	- mel -	- - -
	1	&	2	&
	L	R	L	l
	R	L	R	r

IV. STEPS IN PLACE

15 Man performs 1 step in place on right foot and then brings left foot alongside.

Beginning left foot, woman performs 2 steps in place.

	tjeld ______ .		
	1	&	2
man:	R		
woman:	L		R

REPEAT DANCE FROM BEGINNING

To begin dance again, face CW LOD and assume closed handhold.

Ung Åslaug

Young Åslaug

Formation: Several couples in circle A, facing center of circle. Woman take 1 step forward and join hands with low handhold, forming an inner circle. Men join hands with low handhold, forming an outer circle. The 2 circles should be close together.*

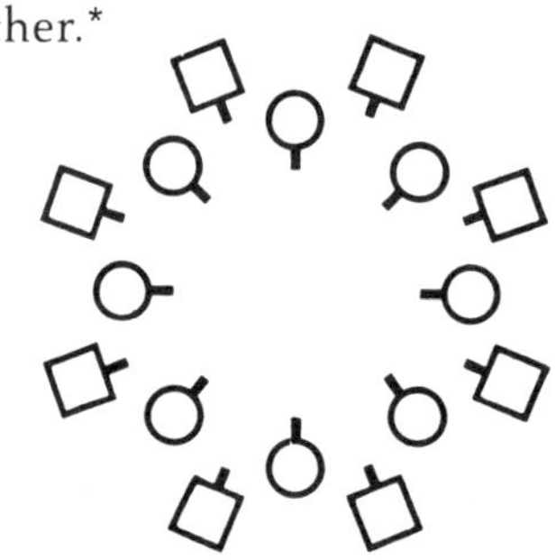

Steps: Light step, open ballad step

* Some Norwegians prefer to begin the dance with everyone in one circle, such that several couples face center of circle with low handhold. These couples then form 2 circles as women step forward in part I.

MUSIC 3/4

Phrase	Ms	Part

I. WOMEN FORWARD AND BACK

A 1-2 Beginning left foot, women dance toward center of circle with 6 light steps. Without releasing handhold, they will, at same time, slowly lift arms forward until hands are almost overhead. Arms are held straight. Women should follow motion of arms with their eyes.

While standing in place, men slowly lift arms in same manner as women.

	Ung Ås-laug	gjekk____		i bjør-ke	skog_____,		så
	1	2	3	1	2	3	
women:	L	R	L	R	L	R	

3-4 Beginning left foot, women dance away from center of circle with 6 light steps, traveling backward. At same time, they lower arms and then release hands, coming to right side of partner in circle. Women should follow motion of arms with eyes. As women come to place on right side of partners, they stand beneath uplifted arms of men.

Men remain standing with arms lifted.

	tir-øygd	frisk____		og	raud ___.	
	1	2	3	1	2	3
women:	L	R	L	R	L	R

II. MEN FORWARD AND BACK

5-6 Beginning left foot, men dance toward center of circle with 6 light steps. Their arms remain lifted with hands joined.

While standing in place, women join hands and slowly lift arms forward until hands are almost overhead. Arms are held straight. Women should follow motion of arms with eyes.

Ho	tyk-ke	mest___	at	al-le	gret___,		der
	1	2	3	1	2	3	
men:	L	R	L	R	L	R	

7-8 Beginning left foot, men dance away from center of circle with 6 light steps, traveling backward. Their arms remain lifted until men come to left side of partners in outer circle. They then release their handhold and assume open-handhold with women.

Women remain standing with arms lifted. As men come to place on left side of partners, women release their handhold and assume open-handhold with men.

	som ho	gle-		da	baud___.	
	1	2	3	1	2	3
men:	L	R	L	R	L	R

III. OPEN BALLAD

B 1-8 Perform 4 open ballad patterns. Following this, lower hands.

Kva	er det	vel___	dei	sy-ta	for___		i
&	1	2	3	1	2	3	
	L	R	L	l	R	r	
	ver-di	vid___		og	stor___?		når
	1	2	3	1	2	3	
	L	R	L	l	R	r	
	so-li	skin___,	og	hjar-ta	brenn___		og
	1	2	3	1	2	3	
	L	R	L	l	R	r	
	kjær leik	-en___		fær	ord___.		
	1	2	3	1	2	3	
	L	R	L	l	R	r	

REPEAT DANCE FROM BEGINNING

To begin dance again, release hands and form original 2 circles as women step forward in part I.

Å kjøre vatten

To Haul Water

Formation: Several couples in circle A, facing CW LOD with closed handhold.

Steps: Closed ballad step, light step

MUSIC 3/4 ______________________________

Phrase	Ms	Part
		I. CLOSED BALLAD
A	1-8	Perform 4 closed ballad patterns.

Å	kjø-re	vat-	ten og	kjø-re	ved	og
	L	R	L	l	R	r
	kjø-re	tøm-mer	o-ver	hei-	a______ .	
	L	R	L	l	R	r
Å	kjø-re	å	dom	kjø-re	vil	je
	L	R	L	l	R	r
	kjø-re	jen-	ta mi	ei-'	a______ .	
	L	R	L	l	R	r

II. WOMAN TURNS CCW; MAN TURNS CW

Release handhold with corner person.

Phrase	Ms	
B	1-2	Man turns CW to face partner and lifts right hand, such that woman can turn beneath it. Beginning left foot, he performs 6 small steps in place.

Beginning left foot, woman makes 1 CCW turn beneath uplifted arms with 6 small steps in place.

Di	rau'e	ro-ser	og di	au-er	blå______ .	
	1	2	3	1	2	3
	L	R	L	R	L	R

Ms	
3-4	Arms remain lifted. Beginning left foot, man makes 1 CW turn beneath upliftied arms with 6 small steps in place.

Woman looks toward man and performs 6 small steps in place.

di	vak-re	jen-ter	hol-der	je	ut-å______ .	
	1	2	3	1	2	3
	L	R	L	R	L	R

III. SIDE STEPS WITH DIPPING ACTION

Man holds partner's left hand from beneath with his right hand, such that her fingers rest on top of his. Man bends right arm and draws woman's left hand across to upper left side of his chest. Both man and woman face center of circle and have some eye contact with each other.

5 Beginning left foot, take 1 step sideways to left, bending at ankle and knee. On second ct, close right foot beside left and transfer weight to both feet. Rise to balls of feet as right foot comes into place. On third ct, lower heels.

helst	når jeg	får,
1	2	3
L	B	B
	rise	lower

6 Repeat action of ms 5, part III.

den	je vil	ha______
1	2	3
L	B	B
	rise	lower

IV. COUPLE TURNS CW

Partners turn to face each other and join hands with position 6.

7-8 Beginning left foot, partners make 1 CW turn in place with 5 light steps. Man then closes right foot beside left. At same time, woman performs a brief curtsey and man acknowledges her.

da	er det	mo-ro-	samt å	le-	ve_______.
	1	2	3	1	2	3
	L	R	L	R	L	R

Note: It is possible to perform this part in a different manner. Beginning left foot, partners make 1 CW turn in place with 3 light running steps. Man then closes right foot beside left. At same time, woman performs a brief curtsey and man acknowledges her.

da	er det	mo-ro	samt å	le-	ve_______.	
	1	2	3	1	2	3
	L	R	L	R		
				curtsey		

REPEAT DANCE FROM BEGINNING

To begin dance again, release handhold and face CW LOD (man makes ½ CCW turn in place). Assume closed handhold.

Figure-Dances

It was hard to tell how many traditional dances lay hidden in the valleys of Norway around the year 1900. Nearly every Norwegian knew that a large collection of old legends and tales had been discovered in some of those valleys by two men with an interest in folklore, the famous Asbjørnsen and Moe, during the early 1800s. But what remained of traditional dances? No one knew for certain, but one young woman from Kristiania, Klara Semb, intended to find out.

Conducting most of her work during the first half of the century, Semb visited several regions in Norway, asking people what they could remember of old dance customs. She encountered a wealth of material. There were couple dances, trio dances, and group dances. There were dances of every temperament, from the lighthearted and flirtatious to the serious and proper. Semb decided to focus on a selection of dances that she called *turdanser.*

"Touring" the Dance Floor

Tur is derived from the French word *tour.*[1] As used in historical descriptions of French dances, a *tour* referred to any action in which one circled about and came back to the same place.[2] The word followed French dances to other countries. Somewhere along the way, however, its meaning changed.

Outside of France, a tour came to mean any figure of a dance, regardless of what sort of action it called for.[3] It was used especially in connection with certain dances that were popular with the upper classes in Europe during the 18th and 19th centuries. Subsequently called "tour dances,"[4] these were performed by a certain number of couples and involved a series of changing geometric formations. Their movements were well defined and followed the music very closely. The quadrille was one example of a tour dance.

Tour dances were done in England and North America. In both places, however, they were more commonly called "set dances."

As used in Norway today, the name *turdanser* may refer to what we would call set dances. Yet it may also refer to other dances that simply resemble set dances in some ways. All have several figures, for instance, that are usually done in the same order and in the same manner each time a given dance is performed. In addition, these figures have been arranged to fit the music, often down to the smallest detail. Semb wrote:

> *It is of no use to be late in a* turdans. *Each musical note has its step, each measure its movement, and all of the claps and stamps must happen at the same second.*[5]

The music is played by a fiddler or other musicians. Some *turdanser* have verses that may be sung by the dancers, but these are rare.

This broader concept of *turdanser* appears to have originated with Semb. I have translated the name as "figure-dances."

It has often been asked why a number precedes the word *tur* in the names of certain figure-dances. Some have suggested that this number may refer to the number of figures or musical phrases. Unfortunately, neither of these suggestions provides a satisfactory explanation in every case. *Tretur* (three-tur) from Fana and *firetur* (four-tur) from Romerike are such exceptions. The first dance has two figures and two musical phrases—not three. The second has three figures and three musical phrases—not four. Consequently, it is difficult to know how these numbers should be interpreted.

Semb brought together a collection of about 40 figure-dances. Directions for these were presented in the several editions of her *Norske Folkedansar* series.

Among Semb's figure-dances are a dozen or so couple dances. Most of these combine turning figures from "old-time" dances (such as the waltz and polka) with two or three other figures to form a short sequence. This sequence is usually danced several times, depending on the length of the music. Couple dances of this kind include *feiar med vals, klappdans, Lottistå, rugen, tantoli, tyrolaren, tyrolarvals,* and *varsovienne.* All were popular during the 19th century. Variants of most of these dances are found in a number of European countries, making it difficult to trace their origins today. *Rugen* is one example. Characterized by an unusual combination of seven and three steps, the dance is found as far south as Yugoslavia.

A Norwegian reel for six persons. Man in front shows off with a sudden twist.

Quite distinct from the couples dances in Semb's collection are Norwegian reels. All are performed by three persons or more. With stamps, claps, quick footwork and rapidly changing figures, they are both challenging and spirited—the kind of dances in which partners can literally kick up their heels.

Reels had their beginnings in Scotland. There they are mentioned in sources dating as far back as the 16th century. The first reels were done by three and four persons. Called *threesome* and *foursome* reels, they took place in a line, such that one or two men danced between two women. After about the year 1800, reels for five and more persons were introduced. These had a variety of formations, including lines, squares, and circles.[6] Reels—particularly threesome and foursome reels—were danced in most Scandinavian countries by the middle of the 19th century.[7]

Some Norwegian reels bear a close resemblance to those of Scotland. *Tretur fra Hordaland,* for instance, has much in common with Scottish threesome reels. These had two fundamental parts. There was a *traveling* figure, during which the dancers wove past each other in a figure-eight pattern, and a *setting* figure, during which the dancers performed a series of nimble steps in place.[8] Both figures are found in *Tretur fra Hordaland.* Curiously enough, threesome reels survived as a living tradition in Norway, but not in Scotland. Our knowledge of Scottish threesome reels comes from dance books that were written during the 19th century.

Other reels have certain mannerisms and figures unique to Norway. The "pinwheel" figure of *seksmannsril,* for example, has no counterpart in Scotland or other Scandinavian countries.

While reels are performed in a robust manner, about a third of the dances in Semb's collection are performed in a very refined and elegant manner. These are 18th and 19th century set dances. Besides graceful movements, the dances are noted for their genteel etiquette. Their figures often begin with sweeping bows and curtseys.

From Folk to Court to Folk Again

Some of the set-dances are descendants of English *country dances.* These began as a folk tradition. Although the early history of the dances is obscure, they are known to have been introduced to the court of Queen Elizabeth about the year 1600, where they were much favored.[9] Professional dancing masters were soon teaching country dances in their classes and composing their own variations. The new court fashion spread to the rest of Europe during the next 200 years. It was called *contredanse anglaise* in France *(contredanse* is generally believed to be a French rendering of country dance), *Contretanz* in Germany, and *contre-danza* in Spain.[10]

Country dances remained popular in some places, including

Norway, until the late 1800s. It is interesting to note that some country dances are found in the earliest-known book of dance instructions from Norway, Sven H. Walcke's *Toure Bog af Engelse og Contra Dandse* (Book for Figures for English and Contra dances), published in Bergen in 1802.

Country dances could take place in a variety of formations. Those in Norway have the so-called "longwayes formation."[11] This called for separate lines of men and women who face each other. Two or three couples at the end of these lines often start the dance and

Den Engelske Dans --"The English Dance." Early 19th century artwork from Denmark captures a country dance in progress. Reprinted on a card for the Art and Industry Exhibit in Copenhagen, 1879. Norsk Folkemuseum.

perform all the figures one time through. One of these couples, the man and woman who stood at the very end of the lines, then continue down the formation, repeating the figures with each of the next couples in succession. Meanwhile, a new group of two or three couples at the end of the two lines could start the dance again.[12] Norwegian country dances with the *longwayes* formation include *feiar* from Sandsvær, *hopsa* from Seierstad, and *sekstur* from Namdal.

A few Norwegian set dances resemble French quadrilles. All the rage in France during the 19th century, quadrilles began as medleys of certain French contre-dances that had been popular during the previous century.[13] They were usually performed by four or eight couples.[14] These couples started in a square formation, but other formations came about with each new figure. Looking at a quadrille from above was somewhat like looking into a kaleidoscope.

One of the more common figures was the *moulinet,* or "mill." Here an even number of dancers traveled in a circle, with right or left hands joined in the center of that circle. The extended arms of the dancers resembled the turning spokes of a mill wheel.[15] Norwegians called this figure *mylnet,* which also means "the mill." It may be seen in *lanciers,* a type of quadrille with five figures.[16] Toward the end of the 19th century, the basic step and turning figure of the waltz were introduced in some quadrilles. These were called "waltz quadrilles."[17] *Firetur* from Romerike and *åttetur* from Asker seem to be two examples.

Semb notated about a dozen other set dances. Although all of these bear some resemblance to dances that were performed by the upper classes in England and France during the 18th and 19th centuries, including country dances and quadrilles, it is hard to make specific connections.

A 19th-Century Ball

Set dances are sometimes described as "court dances" by folk dancers in North America. This name is not entirely accurate. Although some set dances were done by the nobility in the courts of Europe, the same dances were also done by the upper classes in general—even in such faraway places as Hammerfest, a coastal town in the northernmost part of Norway. A ball in Hammerfest was described by a visiting Englishman, William D. Hooker, in his *Notes on Norway* from 1839. He and several of his countrymen volunteered to be the hosts, or "Giest-givers," for this occasion.

> *After breakfast, we went on shore to inspect our ballroom, the late abode of our French friends. The apartment was large and neat, while its elastic flooring made it a capital room for such a purpose. We therefore engaged the Hammerfest Assembly Rooms, and Madame Bang, the landlady, a most notable woman, promised to have everything prepared in a style superior to all that Hammerfest had ever witnessed. . . .*
>
> *I then went back to the* Harriet *to dress for the ball, and was told that every respectable person in Hammerfest had been asked, and that not an individual had refused the invitation; so that we might expect a full room. The hour fixed for meeting was half-past six; we, of course, as the* Giest-givers, *had to be there first, and were shown into an apartment below the salon, where the various eatables were set out. The gentlemen, shortly after, dropped into this room, where we waited to receive them, while the ladies were served with coffee, etc. upstairs. . . . Though some of the gentlemen went into the ballroom before us, still no dancing was attempted until the* Giest-givers *should come.*
>
> *On entering the salon, we found the ladies ranged upon seats against the wall. I may correctly say* ranged, *for the matrons all sat together, then the married damsels, then the younger girls,*

and lastly, those females whose rank did not amount to aristocracy. One corner of the room was filled by a raised platform, and surrounded by a railing, which served as an orchestra for the performers on the violin who, as soon as the Giest-givers *were seen, made quite a flourish of catgut, and presently the whole company were figuring away in a contre-danse.*

It must, however, be remarked that a degree of etiquette is necessary in the opening of the ball. The Giest-giver *on whom this duty devolves, is guilty of an alarming breach of decorum if he does not select first the lady of highest quality in the room. Thus, a lady entitled to the rank of* Frau, *or the wife of some official person, takes the precedence of all the rest, and all the* Fraus *must be handed out before a* Madame *is asked to dance. Waltzes succeeded the first contre-danse, and here the ladies showed great grace and agility. I was somewhat surprised when, after the first dance, glasses of punch, spiced with wine and beer, were handed round, to which we were expected to help our partners, and pledge them. . . .*

During the intervals of dancing, the gentlemen kept walking in and out from the salon, to the smoking, eating and card rooms, where they whiffed their pipes, drank punch, sang, ate sweetmeats, chatted and played at Boston or Whist for very low stakes.

Two quadrilles were attempted, chiefly in compliment to the Giest-givers, *whose performance was by no means such as to raise their character as dancers in the eyes of the Hammerfestian waltzers. One or two* norske *dances followed, through which we contrived to stumble, after many good-natured hints from our partners, and many stupid blunders of our own.*

At midnight, supper was brought in. This consisted of bear and reindeer ham, sandwiches with raw smoked salmon, and the usual accompaniments of punch, liquers, wine, etc. Dancing, and especially waltzing, was again resumed with such vigour, that I soon hardly knew whether it was on my head or heels that I stood, or rather, staggered. At five o'clock, coffee was introduced; at about six, the ladies began to withdraw, and in half an hour or more, the final separation took place; the fun was thus kept up for a whole dozen of hours. No one seemed in the slightest degree tired, except ourselves, and indeed, one of the Giest-givers *was so fatigued that he had to retire and take a nap upon a hard wooden bench. On departing, each individual shook hands with us and thanked us for the pleasure we had conferred.*[18]

It was not always necessary to receive an invitation in order to attend a ball. Any respectable person, including someone without any claim to wealth or status, could attend public balls that were held in some of the more cosmopolitan towns of Europe. The only requirement was the purchasing of a ticket. W. Mattieu Williams appears to have seen one or more balls of this kind while staying in the Norwegian town of Drammen during the autumn of 1859:

I heard many pianos and a good deal of singing in the houses, and passed some public rooms where there were many dancers.[19]

A more complete description of a public ball, held in Kristiania, may be found in the next chapter.

"Student Society Ball." A program for a ball held in Kristiania, 1899. Handskriftsamlingen, Universitetsbiblioteket, Oslo.

Ballroom dances, such as those mentioned in the Hammerfest description, were also adopted by certain well-to-do people who lived on the outskirts of the towns and in the countryside. Finding it difficult to attend balls in the towns when their primary means of transportation was the horse and carriage, these people would come together now and then with a select number of friends for a polite evening of conversation, refreshments, and dancing. Such a gathering was also called a ball, even though it usually took place in a private home.[20] Those who attended might include wealthy merchants, members of the clergy, local magistrates, military officers, farmers with large estates, and the wives of each. Older sons and daughters would often be invited as well. Due to their relative isolation, these people led a rather provincial way of life, sometimes holding onto certain fashions for many years after these had gone out of style elsewhere. They often contrived their own regional variations of popular dances. It was among them that Semb found some of her figure-dances.

A trio dance from Fana in the Hordaland province.

More than a few figure-dances have no clear relationship to dances from other parts of Europe. Especially curious is *krossadans med seks* (cross dance with six people) from the Sogn district. Four men and two women perform this dance instead of evenly matched couples. From an interesting weaving figure to an almost Hindu-like bow at its conclusion, the dance has several unique features, possibly Norwegian in origin.

Semb was not the only person to go in search of figure-dances. Among others, Egil Bakka has collected several couple dances, reels, and set dances. He gave directions for these in his book, *Danse, danse, lett ut på foten* (Dance, Dance, Light on the Feet), published in 1979.

Perhaps the most remarkable feature of figure-dances in Norway is their diversity. One finds a broad selection of dances from different locations and social environments. Those that were done by the upper classes are of special interest. Although influenced by trends from abroad, these people often interpreted these trends in distinctive ways, especially those who lived outside of the cities. Their dances help to illuminate an important side of Norwegian culture that is often overlooked.

Notes

1 *Riksmals ordboken* (Oslo, 1977). s.v. "tur."

2 Albert Zorn, *Grammatik der Tanzkunst* (Leipzig: 1887)-*Grammar of the Art of Dancing,* trans. Benjamin P. Coates, ed. Alfonso J. Sheafe (New York: Burt Franklin, 1905, reprint 1970), p. 181.

The *tour de main* of a quadrille, for instance, was a figure in which a couple joined hands and circled about, making one rotation. Melusine Wood, *Advanced Historical Dances* (London: C. W. Beaumont, 1960), pp. 93-94.

3 Zorn, *Grammatik,* p. 181.

4 Franz M. Bohme, for instance, described the *ecossaise* as a *Tourentanz* in his book, *Geschichte des Tanzes in Deutschland* from 1886.

5 Klara Semb, *Norske Folkedansar II* (Oslo: Noregs Boklag, 1956), p. 162.

6 J. F. Flett and T. M. Flett, *Traditional Dancing in Scotland* (London: Routledge & Kegan Paul Ltd., 1964), pp. 132-33.

7 Yngvar Heikel, "Folkdans," *Finlands Svenska Folkediktning,* vol. 6B (Helsingfors: Mercators Tryckeri, 1938), see "Engelska-danser," pp. 264-281; Tobias Norlind, *Dansens Historia: Med särskild hänsyn till dansen i Sverige* (Stockholm: Nordisk Rotogravyr, 1941), p. 118; Svenska Ungdomsringen För Bygedekultur, *Svenska Folkdanser* (Falun: Falu Nya Boktryckeri Aktiebolag, 1975), see "Tretur," pp. 112-113.

8 Flett and Flett, *Traditional Dancing,* pp. 132-33.

9 George S. Emmerson, *A Social History of Scottish Dance* (Montreal and London: McGill-Queens University Press, 1972), pp. 61-62.

10 A. P. Oppé and Cecil J. Sharp, *The Dance: An Historical Survey of Dancing in Europe* (London: Halton & Truscott Smith, Ltd., 1924), pp. 24-30.

11 "Longwayes," first description by John Playford, *The English Dancing Master* (London: Thomas Harper, 1651).

12 Thomas Wilson, *English Country Dancing* (London: Sherwood, Nelly, and Jones, 1820), pp. 151-68, 183-85.

13 Philip J. S. Richardson, *Social Dances of the Nineteenth Century in England* (London: Herbert Jenkins Ltd., 1960), pp. 22-28.

14 Zorn, *Grammatik,* p. 183.

15 Wood, *Advanced Historical Dances,* p. 94.

16 Lanciers, known in England as lancers. A lancers was a quadrille with five figures. See Oppé and Sharp, *The Dance,* p. 30

17 P. Valleau Cartier, *Waltz Instructor and Ballroom Guide* (New York: De Witt, Publisher, 1882), p. 21-23.

18 William D. Hooker, *Notes on Norway,* (Glasgow: George Richardson, 1839), pp. 24-28.

19 W. Mattieu Williams, *Through Norway with a Knapsack* (London: Smith, Elder and Co., 1859), p. 322.

20 Thora Skolmen, "Leik og Dans," *Folkeminne Tidskrift* (Skien: Norsk folkekultursforlag, 1935), p. 90; see also Olav Bø, "Livet i laget," *Gilde og Giestebod,* ed. Halvor Landsverk (Oslo: Det Norske Samlaget, 1976), p. 163.

Feiar fra vestlandet

Feiar from Western Norway

Formation: 6 or more couples of an even number in circle B. Couples are numbered off in ones and twos. Partners face each other with light handhold extended to side (man's right, woman's left). Free hand is on hip, thumb forward.

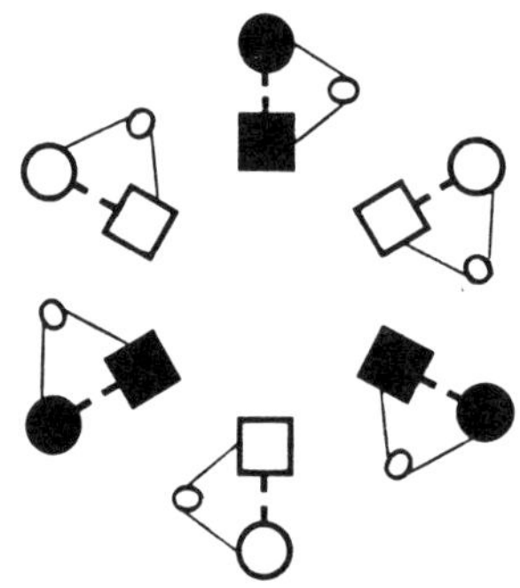

Steps: Arm–swing turn, change step, step-hop

MUSIC 2/4

Phrase	Ms	Part
		I. ARM-SWING TURN; STEPS FORWARD
A	1-3	With man beginning left foot and woman right, perform 3 arm-swing turns in CCW LOD.
	4	With man beginning right foot and woman left, take 2 steps forward in CCW LOD with light stamps on both steps; turn to face partner on second step.
	5-8	Take hands with light handhold extended to side (man's left hand, woman's right). Free hand is on hip, thumb forward. Repeat action of ms 1-4, part I in opposite direction with opposite footwork.
		II. STAR CW AND CCW
		No. 2 couples turn (man CW, woman CCW) to face no. 1 couples:

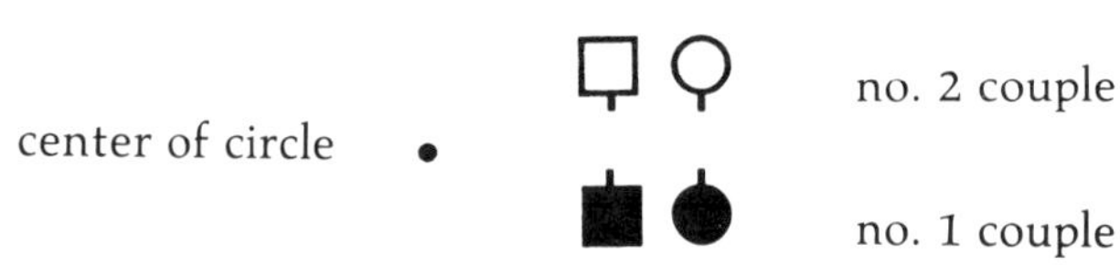

B 1-4 No. 2 and no. 1 couples clap hands on ct 1 and quickly form a "star." No. 2 man and no. 1 woman extend right arms and interlock right hands by grasping around base of thumb area and onto upper surface of hand, such that thumbs are uppermost. No. 1 man and no. 2 woman do same, joining hands close below the other handhold. Left hand is on hip, thumb forward.

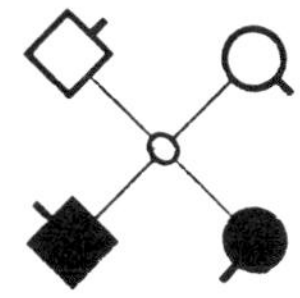

Beginning left foot on ct 1, perform 8 step-hops in CW LOD in star formation. On last step-hop, release handhold and make ½ CW turn on supporting right foot.

5-8 Clap hands together on ct 1 and form a star by joining left hands with thumb grasp.

Beginning left foot on ct 1, perform 8 step-hops in CCW LOD. Star should rotate about same speed as first time so that everyone will end up in original positions.

III. TWOS MAKE ARCHES; ONES GO UNDER

A 1-4 No. 2 couples make an arch with inside hands (man's left hand, woman's right) such that no. 1 couples will be able to pass beneath. No. 1 partners face each other with light handhold extended to side (man's right hand, woman's left). Free hand for both couples is on hip, thumb forward.

center of circle •

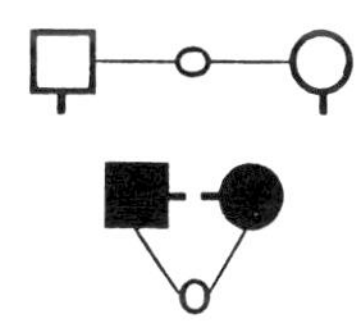

no. 2 couple

no. 1 couple

No. 1 couple repeats action of ms 1-4, part I, bending down and passing beneath arch on first arm-swing turn.

With man beginning left foot and woman right, no. 2 couple performs 1 change step toward no. 1 couple in CW LOD on cts 1-&-2, allowing no. 1 couple to pass beneath arch. No. 2 partners then pivot on supporting foot (man left foot, woman right) and make 3/4 turn (man CCW, woman CW) such that they conclude back-to-back. As partners turn, they change their handhold. Man quickly takes woman's left hand with his right, swinging it down and forward as he does so:

center of circle •

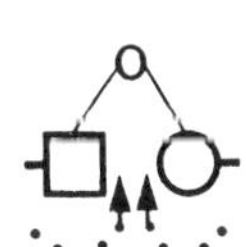

no. 2 couple

From here no. 2 couple performs 2 arm-swing turns (the first is done back-to-back). With man beginning right foot and woman left, no. 2 partners take 2 walking steps forward in CCW LOD with light stamps on both steps; they turn to face each other on second step.

5-8		All couples repeat action of ms 5-8, part I.

IV. STAR CW AND CCW

No. 2 couple turns to face no. 1 couple as before. Since no. 1 couples have now progressed forward (as a result of going under arches), star formation will be done with new pairs of couples.

B	1-8	Repeat action of ms 1-8, part II.

* * *

Repeat parts III and II for remainder of music. Note that this is a progressive dance in which couples are always changing their relative positions.

Feiar med vals

Feiar with Waltz
Ringerike, Buskerud

Formation: Several couples in circle B with position 2.

Steps: Arm-swing turn, change step, step-hop, waltz

MUSIC 3/4

Phrase Ms Part

I. WALTZ

A 1-16 With man beginning left foot and woman right, couples waltz for 16 ms, making 8 complete turns.

MUSIC 2/4

II. ARM-SWING TURN; STEPS FORWARD

Partners face each other with light handhold extended to side (man's right hand, woman's left). Free hand is on hip, thumb forward.

B 1-3 With man beginning left foot and woman right, perform 3 arm-swing turns in CCW LOD.

4 With man beginning right foot and woman left, take 2 steps forward in CCW LOD with light stamps on both steps; turn to face partner on second step.

5-8 Take hands with light handhold extended to side (man's left hand, woman's right). Free hand is on hip, thumb forward. Repeat action of ms 1-4, part II in opposite direction with opposite footwork.

III. POSITION 2; INTRODUCTORY TURN

C 1 Release handhold.

Beginning left foot, man performs 1 change step almost in place, turning a little CCW as he does so. At same time, he places his right hand around the waist of his partner and draws her forward and in front of him to assume position 2. Man may stamp on first step.

Beginning right foot, woman dances in front of man and turns to face him with 1 change step. By end of ms, partners should be in position 2. Man should be approximately facing CCW LOD.

2 Couple makes about ¼ CW turn with 1 change step. Man should conclude facing away from center of circle, woman toward.

3-4 With man beginning left foot and woman right, make 1 CW turn with 2 change steps.

IV. COUPLE TURNS CW

5-7 With man beginning left foot and woman right, make about 3½ CW turns with 6 smooth step-hops. Extra half turn means that couple must rotate more quickly than usual on last 2 step-hops. Man should conclude facing toward center of circle, woman away.

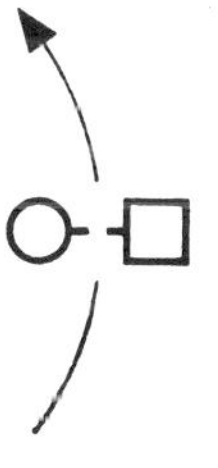

V. LIFT

Man securely places hands on either side of woman's waist. Woman places both hands firmly on top of man's shoulders. Man and woman place weight over both feet.

8 On ct 1, man lifts woman by waist, using good flexion in knees, and makes ½ CW turn with her in air. On ct 2, he puts woman down on other side.

Woman prepares for lift by bending knees a little and then by jumping vertically on ct 1. As woman jumps, she also presses downward on man's shoulders, partially supporting her weight. Woman lands with flexion in ankles and knees.

Note: This lift is sometimes omitted by dancers in Norway. They perform 8 step-hops instead of 6 and make 4 CW turns.

Figaro der alle danser

Figaro Where Everyone Dances

Formation: 8 couples arranged in 4 groups. In each group, one couple stands across from another couple. Man is on left side of partner. Those who face away from front of room are no. 1 couples. They are pictured in black. Those who face toward front are no. 2 couples. Note that room is divided into side A and side B.

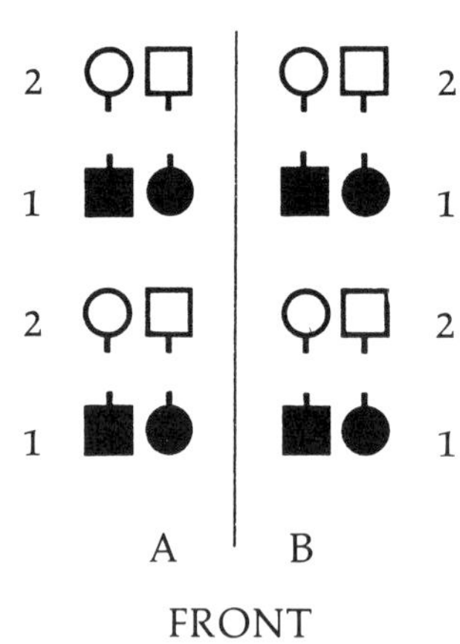

Steps: Change step (¾ rhythm, first version), touch step, waltz

MUSIC 3/4

Phrase Ms Part

I. CIRCLE CW AND CCW; ACKNOWLEDGEMENT

A 1-3 No. 1 and no. 2 couples join hands in circle with light handhold.

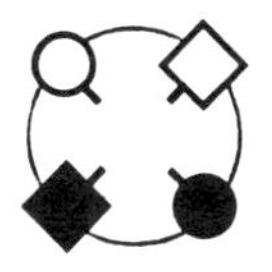

Beginning left foot, perform 3 touch steps in CW LOD.

4 On ct 1, turn and step backwards out of circle with right foot, such that left foot remains in place and extends. Perform an acknowledgement toward center of circle. Hands remain where they are in ring formation, such that arms will extend somewhat when stepping backward.

5-7 Beginning left foot, perform 3 touch steps in CCW LOD.

8 Repeat action of ms 4.

II. TRAVEL TO OTHER SIDE

Release handhold. Man offers woman his right arm in "escort position." His elbow is bent, such that woman can rest her hand on man's arm. Man places left hand low on hip, fingers forward. Woman holds skirt to side with right hand. As partners assume this position, they briefly acknowledge each other.

No. 1 and no. 2 couples on side A will now exchange places with their counterparts on side B and vice versa.

9-10 Beginning left foot, dance 2 touch steps. Each couple will move as shown below:

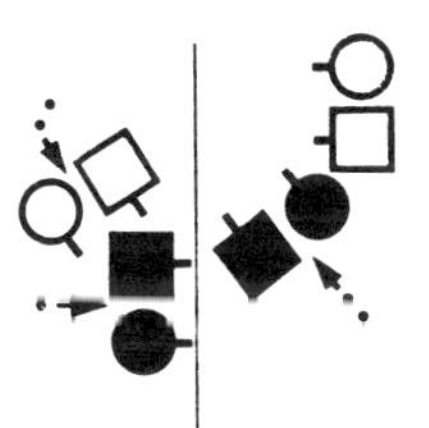

FRONT

Note that no. 1 couples progress in front of no. 2 couples.

11-13 Beginning left foot, dance 3 touch steps. No. 1 couples lead no. 2 couples to opposite side.

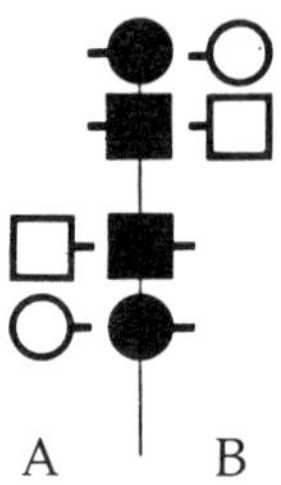

A B

FRONT

14-16 Beginning right foot, dance 3 touch steps. Each couple will move as shown below:

A B

FRONT

Note that no. 1 couple from side B curves in front of no. 2 couple, such that no. 1 man leads with his left shoulder. Similarly, no. 2 couple from side A curves in front of no. 1 couple, such that no. 2 man leads with his left shoulder.

By end of last ms, exchange should be complete. As at beginning of dance, no. 1 couples face away from front of room and no. 2 couples toward front:

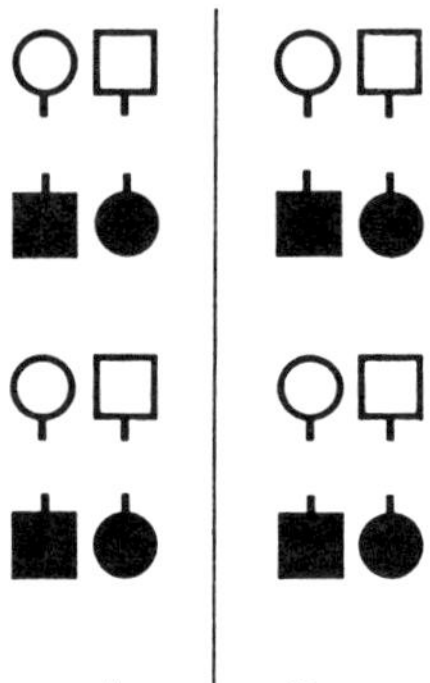

A B

FRONT

III. CHAIN

Release escort position. Partners turn on right foot to face each other (man ¼ CW turn, woman ¼ CCW turn).

B 1-7 Take right hands with partner and chain past 4 persons, counting partner as no. 1. Beginning left foot, dance 2 change steps per person until you meet fourth person; dance 1 change step with him or her. Conclude in starting positions.

8 Beginning right foot, dance 1 change step, moving into position for traveling figure (no. 1 couples face away from front of room, no. 2 couples toward front). Partners acknowledge each other and assume escort position.

IV. TRAVEL TO OTHER SIDE

9-16 Couples from side A have now become couples on side B and vice versa. Repeat action of ms 1-8, part II. It may be helpful to remember that no. 1 couples always dance in front of no. 2 couples. You will return to same place as at beginning of dance.

V. STAR CW AND CCW

C 1-7 Each group of no. 1 and no. 2 couples will form a "star." Face CW LOD. Men lightly join right hands across. Women do same, taking hands directly below men's handhold. Man's left hand is low on hip, fingers forward. Woman holds skirt to side with right hand.

Beginning left foot, perform 7 change steps in CW LOD. Star will rotate about 1½ times.

8 Release handhold. On ct 1, step backwards out of ring with right foot, such that left foot remains in place and extends. Perform a brief bow toward center of group. Man places his right hand low on hip, fingers forward, and gracefully swings his left arm in front of his midsection. Woman holds skirt to either side with both hands.

9-15 Couples in each group face CCW and form a star, this time with left hands lightly joined across.

Beginning left foot, perform 7 change steps in CCW LOD.

Star will rotate about 1½ times to original place.

16 Repeat action of ms 8, part V with following change: Man places his *left* hand low on hip, fingers forward, and swings his *right* arm in front of midsection. On ct 3, man shifts weight onto left foot in preparation for waltz.

VI. WALTZ

D 1-10 Partners come together with 1 waltz step to assume position 1. Man begins right foot and dances toward woman; woman begins left foot and dances mostly in place. Couples then waltz for next 9 ms, such that each group of no. 1 and no. 2 couples makes its own small circle to travel in. Couples should conclude approximately where they began to waltz.

11-14 Couples continue to waltz for another 4 ms, traveling around small circles. During this time, all couples who have progressed toward either end of room will waltz toward corner of room that is nearest them.

15-16 Partners open away from each other and assume escort position. They should conclude in following formation:

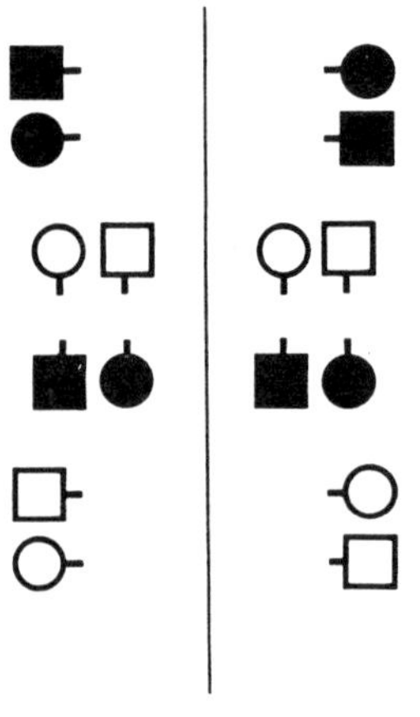

FRONT

Each couple in center of room will now form a new group with couple across from them. Couples in each of 4 corners will now stand in place and be inactive for a while.

* * *

Center couples repeat dance from beginning.

During last 6 ms of waltz, 4 new groups will be formed by joining center couples with inactive couples, who have been waiting in each corner. Beginning right foot, inactive couples perform 3 touch steps as they come forward to meet center couples:

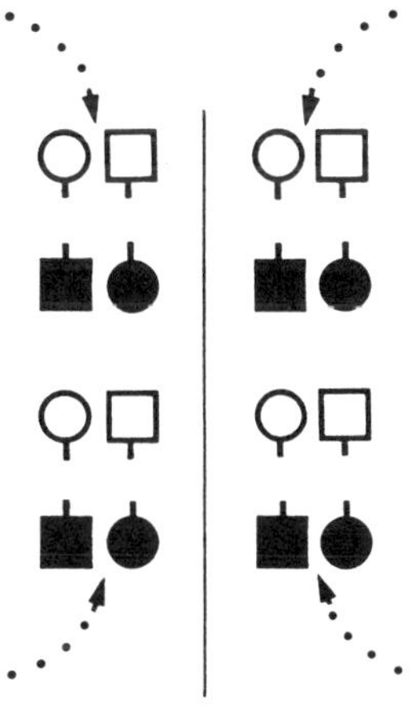

FRONT

Note: As inactive couples come into new groups, they change roles. No. 1 couples become no. 2 couples and vice versa.

REPEAT DANCE FROM BEGINNING

VARIATION

You may have noticed that no. 1 couples progress away from front of room and no. 2 couples toward front:

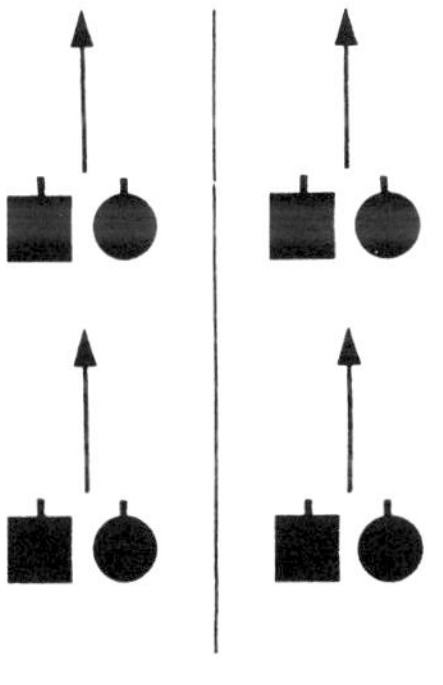

FRONT

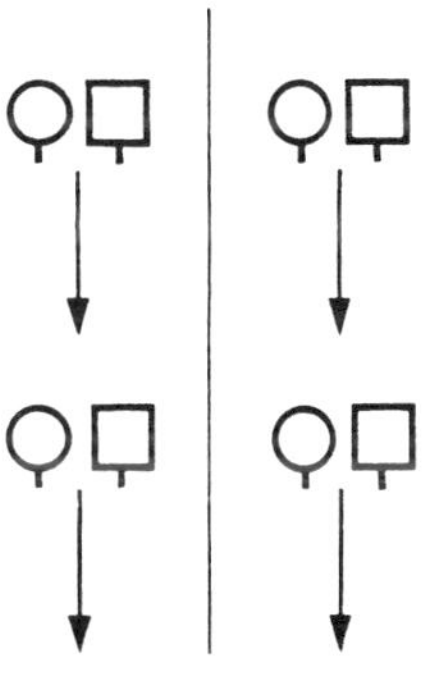

FRONT

Once a couple reaches end of formation, they must turn around and come back again with a different role. This "recycling" of dancers makes it possible to rotate 2 additional couples in on one end of formation:

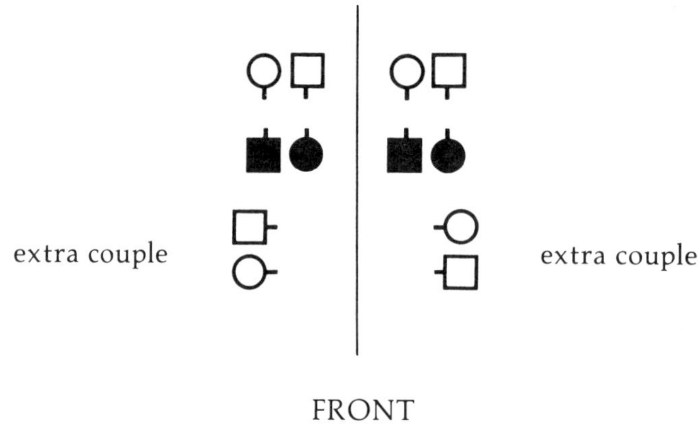

These 2 additional couples are inactive for first performance of dance and then are active during next performance. They become no. 1 couples, joining with no. 2 couples who come toward them at end of waltz.

As the 2 new couples enter dance, another 2 couples on opposite end of formation will progress out of dance for time being:

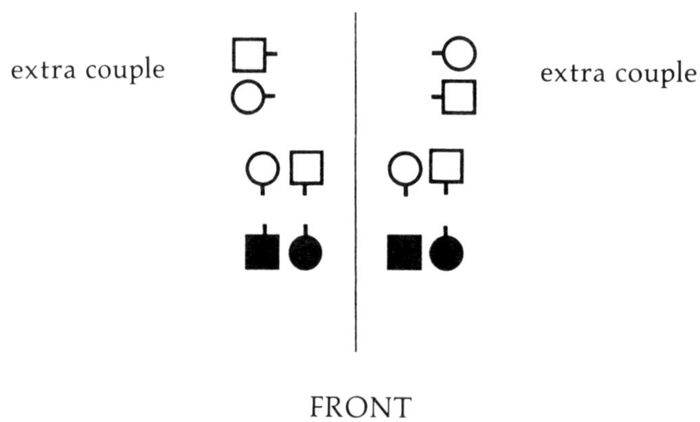

In this variation there are always 2 rotating couples on one end of formation who are inactive for 1 performance of dance. The advantage of this variation is that 6 or more couples of any even number can take part.

Firetur

Landvik, Aust Agder

Formation: 4 or more couples in circle A, facing center of circle with light handhold.

Steps: Gesture step, light step, reel step, step-hop

MUSIC 2/4 ______________________________

Phrase	Ms	Part	
		I.	**CIRCLE CW AND CCW**
A	1-4		Beginning left foot, perform 8 step-hops in CW LOD. On last step-hop, make ½ CW turn.
	5-8		Beginning left foot, perform 8 step-hops in CCW LOD.
		II.	**GESTURE STEP; REEL** Release handhold. Turn to face corner person. Man makes ½ CCW turn on supporting right foot. (Note: when this figure is repeated later in the dance, woman may need to make ½ CW turn to face corner person.) Man folds arms above chest and parallel to ground. Woman holds skirt to either side.
B	1		Beginning left foot, perform 1 gesture step pattern to right.
	2		Beginning right foot, perform 1 gesture step pattern to left.
	3-4		Repeat action of ms 1-2, part II. With hopping action at end of last gesture step, make ½ CW turn on supporting right foot and face partner.
	5-8		Beginning left foot, perform 8 reel steps in place while facing partner. (Man's arms should still be folded and woman's hands are still on skirt.) With last step, make ½ CCW turn on supporting right foot to face corner person.
		III.	**FIGURE EIGHT; INFORMAL GREETING**
C	1-2		Extend right arm toward corner person and grasp around his or her elbow. Beginning left foot, make ½ CW turn around corner person with 3 light steps. Take a fourth light step and at same time, release elbow hold and move toward your partner, who should be ahead of you.
	3-4		Extend left arm toward partner and grasp around his or her elbow. Beginning left foot, make ½ CCW turn around partner with 3 light steps. Take a fourth step, and at same time, release elbow-hold and move toward corner person, who should be ahead of you.

With completion of ms 4, one "figure eight" has been completed. Floor patterns for man and woman are shown below.

5-6 Repeat action of ms 1-2, part III.

7 Beginning left foot, move toward partner with 2 light steps; stop a little distance in front of partner.

8 Man folds arms high above chest and parallel to ground. Woman holds skirt to side. Couple performs informal greeting on last 2 extended cts in ms. Man bends forward from waist on first ct and comes up on second. Woman lowers on first ct and comes up on second.

MUSIC 3/4

IV. WALTZ; ACKNOWLEDGEMENT

D 1-10 With man beginning right foot and woman left, partners come together with 1 waltz step to assume position 2 and make ¼ CW turn, such that man concludes facing away from center of circle, woman toward. Partners then waltz as usual for 9 ms, making 4½ CW turns.

11 Partners now dance 1 waltz step almost in place as they turn into their positions in ring A. As this happens, they open away from each other. Woman slides her left hand along man's right arm to his hand. Man holds arm out to steady her.

12 Woman holds skirt to side with right hand. She performs an acknowledgement toward partner, placing right foot behind left.

Man places left hand low on hip, fingers forward. He performs an acknowledgement toward partner, placing left foot behind right.

Note: Some melodies have 16 measures for this part. For such melodies, waltz through 14th ms, open away from partner on 15th ms, and acknowledge partner on 16th ms.

V. CW AND CCW TURN

Partners assume following position:

A 1-4 Beginning left foot, partners make 2 CW turns with 8 step-hops.

5-8 Partners assume following position:

Beginning left foot, partners make 2 CCW turns with 8 step-hops.

VI. REFRAIN

BCD Repeat action of all ms in parts II, III, and IV.

VII. CIRCLE CW AND CCW; ONE ARM UP

Upper body faces approximately toward center of circle; hips are angled toward CW LOD.
Assume following position:

A 1-4 Beginning left foot, perform 8 step-hops in CW LOD. On last step-hop, make ½ CW turn.

5-8 Clap hands together in front of chest on ct 1 and assume following position:

Beginning left foot, perform 8 step-hops in CCW LOD

VIII. REFRAIN

BCD Repeat action of all ms. in parts II, III, and IV.

IX. CIRCLE CW AND CCW; BOTH ARMS UP

Upper body faces approximately toward center of circle; hips are angled toward CW LOD. Assume following position:

A 1-4 Beginning left foot, perform 8 step-hops in CW LOD. On last step-hop, make ½ CW turn.

5-8 Beginning left foot, perform 8 step-hops in CCW LOD.

X. REFRAIN

BCD Repeat action of all ms in parts II, III, and IV.

XI. CIRCLE CW AND CCW; CLAPPING

Face CW LOD.

A 1-4 Beginning with ct 1, clap hands together on each ct of music in special manner: hold right hand in front of chest with palm facing up and clap left hand onto it; reverse procedure for next clap. (Action resembles flipping a piece of dough back and forth.)

Beginning left foot, perform 8 step-hops in CW LOD. On last step-hop, make ½ CW turn.

5-8 Beginning left foot, perform 8 step-hops in CCW LOD; continue clapping.

XII. REFRAIN

BCD Repeat action of all ms in parts II, III, and IV.

XIII. CW AND CCW TURN

A 1-8 Repeat action of ms 1-8, part V.

Firetur

Romerike, Akershus

Formation: 2 couples in circle A, facing center of circle with light handhold.

Steps: Touch step, waltz

MUSIC 3/4

Phrase	Ms	Part	
		I.	**CIRCLE CW AND CCW; ACKNOWLEDGEMENT**
A	1-7		Beginning left foot, perform 7 touch steps in CW LOD.
	8		On ct 1, turn and step backwards out of circle with right foot, such that left foot remains in place and extends. Perform an acknowledgement toward center of circle. Hands should remain where they are in ring formation, such that arms will extend somewhat when stepping backward.

Circle in clockwise line of direction

Acknowledgement

Phrase	Ms	Part	
	9-15		Beginning left foot, perform 7 touch steps in CCW LOD.
	16		Repeat action of ms 8.
		II.	**CHAIN**
			Release handhold. Partners turn to face each other (man ¼ CW turn, woman ¼ CCW turn).

B 1-16 Take right hands with partner and chain past 8 persons, counting partner as no. 1. Beginning left foot, dance forward in tight circle with 2 touch steps per person. Men move in CCW LOD, women in CW LOD. Stop in front of partner.

Chain figure

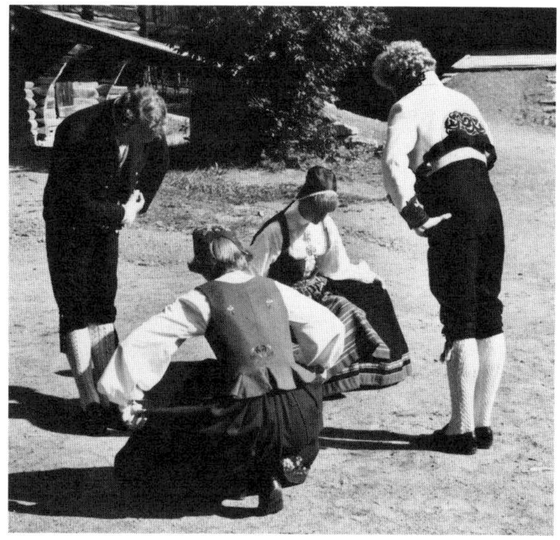

Formal greeting

III. FORMAL GREETING

See instructions for formal greeting.

C 1-4 Man bows forward in 6 cts and rises in 6 cts. Woman lowers with formal curtsey in 6 cts and rises in 6 cts.

IV. WALTZ; ACKNOWLEDGEMENT

5-14 Partners come together with 1 waltz step to assume position 2. Man begins right foot and dances toward woman; woman begins left foot and dances mostly in place. Couples then waltz for the remaining 9 ms, making 4½ CW turns in overall CCW LOD.

15 Partners now dance 1 waltz step almost in place as they turn into their original starting positions in circle A. While this happens, they open away from each other. Woman slides left hand along man's right arm to his hand. Man holds arm out to steady her.

16 Woman holds skirt to side with right hand. She performs an acknowledgement toward partner, placing right foot behind left.

Man places left hand low on hip, fingers forward. He performs an acknowledgement toward partner, placing left foot behind right.

REPEAT DANCE FROM BEGINNING

Krossadans med seks

Cross-dance with Six Persons
Brekke, Sogn

Formation: 2 men and 4 women in circle, facing center of circle with light handhold:

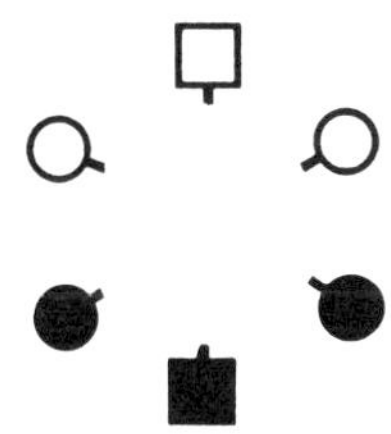

Note than men and women are divided into two groups by color.

Steps: Light running step, light step

MUSIC 2/4

Phrase Ms Part

I. CIRCLE CW AND CCW

A 1-4 Beginning left foot, perform 8 light steps in CW LOD. On last step, pivot on supporting right foot and make ½ CW turn.

5-8 Beginning left foot, perform 8 light steps in CCW LOD.

II. TURN CIRCLE INSIDE OUT; REVERSE

At this point rhythm changes. There will be a musical phrase with 20 cts, which is then repeated.

B Without releasing hands, man in black group makes ½ CCW turn beneath his left arm and leads everyone in CCW LOD, such that circle is turned inside out. Beginning left foot, perform this with 20 light running steps. Note that everyone continues to travel in CCW LOD after ciircle is turned inside out.

b Without releasing hands, man in black group makes ½ CCW turn beneath his left arm and leads everyone in CW LOD, such that circle is turned "outside in." Beginning left foot, perform this with 20 light running steps. Note that everyone continues to travel in CW LOD after circle is turned "outside in." Everyone should conclude in original starting positions.

III. CIRCLE CW AND CCW

A 1-4 Repeat action of ms 1-4, part I.

5-8 Repeat action of ms 5-8, part I, but with one change. On last 2 steps, release hands and form 2 lines as shown below:

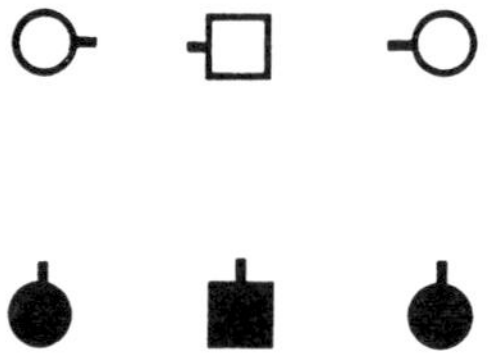

Man in white group folds arms above chest and parallel to ground. Women in white group place hands on hips, thumbs forward. (Women may also hold skirt to either side.) Dancers in black group relax arms to either side.

IV. WEAVE

Dancers in white group stand in place, leaving some space in between themselves. Dancers in black group will now weave in and out of white group as follows:

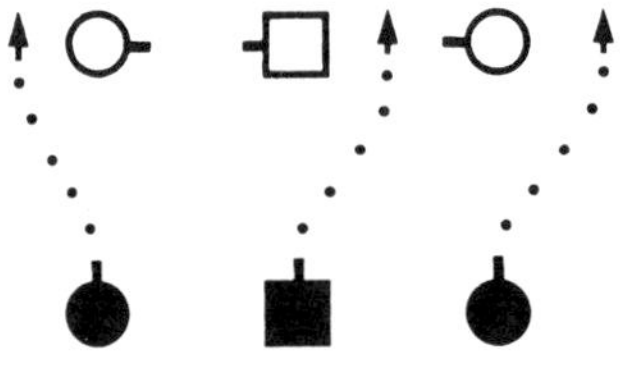

B Beginning left foot, each person in black group performs about 4 light running steps forward, moving toward back side of person in front of him or her:

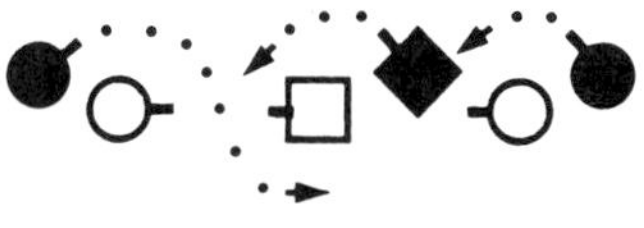

Each person in black group then goes around this stationary person and weaves in and out of other persons in white group with about 32 light running steps.

This weave should be done such that each woman in black group will go past all 3 persons in white group, loop around end person, and then come back. They conclude weave by going completely around same person that they went around at beginning of weave. Floor pattern for man in black group is somewhat different and is shown below:

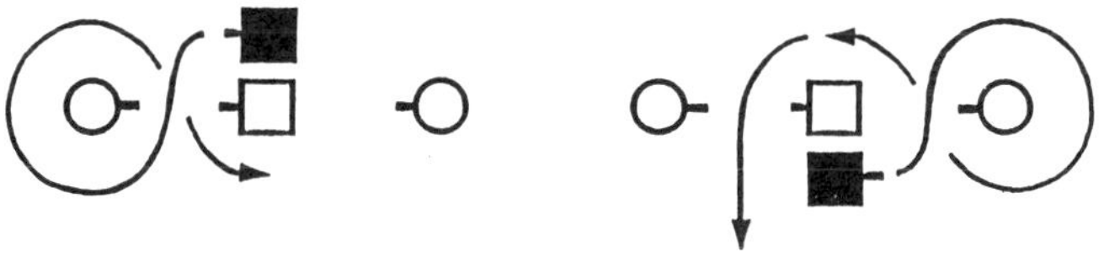

Dancers in black group return to their original starting positions with about 4 light running steps, making ½ CCW turn as they do so.

V. CIRCLE CW AND CCW
Join hands in circle with light handhold.

A 1-4 Repeat action of ms 1-4, part I.

5-8 Repeat action of ms 5-8, part I, but with one change. On last 2 steps, release hands and form 2 lines as shown below:

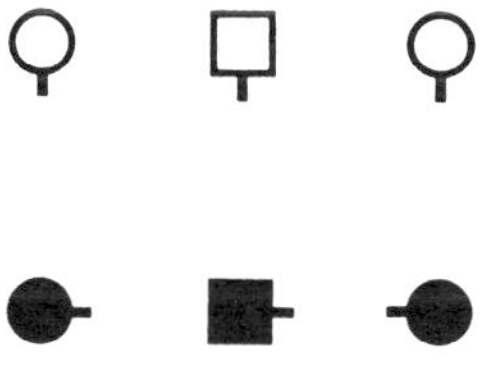

Man in black group folds arms above chest and parallel to ground. Women in black group place hands on hips, thumbs forward. (Women may also hold skirt to either side.) Dancers in white group relax arms to either side.

VI. WEAVE
B Repeat action of part IV. Dancers in white group perform weave.

VII. CIRCLE CW AND CCW
Form a circle by placing hands on shoulders of person on either side of you.

A 1-8 Repeat action of ms 1-8, part I.

VIII. CIRCLE CCW
B On ct 1, perform a small jump onto both feet. Beginning right foot on ct 2, perform 15 light running steps in CCW LOD.

Turn to face center of circle. At same time, lift hands above head and clap them together. Keep hands together and lower them with extended arms in front of body. Bow toward center of circle.

Rugen

Formation:		Several couples in circle A with position 1. Man faces CCW LOD, woman CW LOD.
Steps:		Step-hop
MUSIC 2/4		

Phrase	Ms	Part
		I. SIDE STEPS TOWARD CENTER; AWAY
A	1-2	With man beginning left foot and woman right, perform 6 steps toward center of circle on cts 1-&-2-&-1-&. Steps travel sideways (step-close, step-close, step-close). Following this, take one more step toward center of circle and stamp on that step (man's left foot, woman's right). Pause on upbeat.
	3-4	Repeat action of ms 1-2 in opposite direction with opposite footwork.
		II. INTRODUCTORY TURN
	5	With man beginning left foot and woman right, partners perform 3 steps almost in place on cts 1-&-2, turning a little CCW as they do so. Stamp on third step.
	6	With man beginning right foot and woman right, partners perform 3 steps almost in place on cts 1-&-2, turning a little CW as they do so. Stamp on third step.
		III. COUPLE TURNS CW
	7-8	With man beginning left foot and woman right, make 1 or 2 CW turns with 4 step-hops. Footwork is interlocking, such that man and woman step between partner's feet with right foot. Partners should conclude in original starting positions with man facing CCW LOD and woman CW LOD.
	9-12	Repeat action of ms 5-8.

Sandsværril

Reel from Sandsvær
Sandsvær, Buskerud

Formation: 4 or more couples in circle A, facing center of circle with light handhold.

Steps: Change steps, light step, various steps for men (described below).

MUSIC 2/4 ______

Phrase Ms Part

I. CIRCLE CW AND CCW

A 1-8 Beginning left foot, perform 14 light steps in CW LOD. Following this, stamp 3 times in place with left-right-left, cts 1-&-2.

9-16 Beginning right foot, turn and perform 14 light steps in CCW LOD. Following this, stamp 3 times in place with right-left-right, cts 1-&-2.

II. TRAVEL CW AND CCW; MAN SHOWS OFF

Release handhold and turn on supporting right foot to face partner.

B 1-8 Beginning left foot, woman dances backward with 8 change steps in CCW LOD. She holds skirt to either side. Woman interacts with man, watching his dancing as he tries to impress her. On last step, make ½ CCW turn to face corner person.

Beginning left foot, man dances forward with 8 change steps. His arms hang naturally to either side. Man can, however, substitute change steps with various other steps when he wishes to do so. Three such steps are described below:

Step One

Step forward with either foot on ct 1. Slap side of other foot with nearest hand on ct 2. (Your right hand, for instance, would slap side of right shoe, making a sound. Your foot should come up to your hand, **do not** bend or tilt over in order to reach foot.) Next step pattern begins immediately on foot you slapped.

Step Two

Beginning either foot, make 1 CW or CCW turn with 2 change steps.

Step Three

Beginning either foot, make 1 CW or CCW turn with 1 change step.

Man's arms may be placed in different positions: let arms hang naturally to either side; put thumbs in armholes of vest; or fold arms above chest and parallel to ground. On last step, man makes ½ CW turn to face corner person.

9-16 Beginning left foot, repeat action of ms 1-8, part II, now facing corner person and dancing in CW LOD. Woman dances backward, man forward.

III. "HANDLESS CHAIN"

A "handless chain" is a variation of the chain figure. Dancers weave in and out of circle as in the usual chain, but they do not take hands. Man's arms are relaxed by his sides, swinging naturally with his movements. Woman holds skirt to either side.

Remain facing corner person.

C 1-6 Without taking hands, chain past 6 persons, counting corner person as number 1. Beginning left foot, dance 1 change step per person. Men should travel in CW LOD, women CCW LOD.

7 Upon passing sixth person (who will become your new partner), everyone will make ½ CCW turn around that person with 1 change step, such that you are face-to-face as you turn around. Besides facing new partner, you should conclude in a circle formation, such that man faces CCW LOD, woman CW LOD:

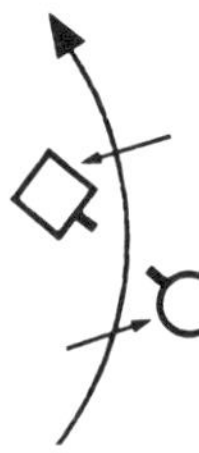

passing sixth person

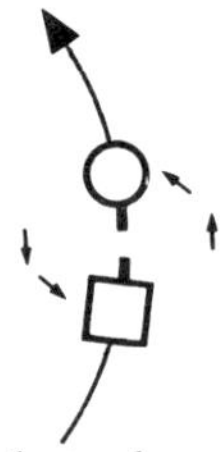

around sixth person

8 Stamp 3 times in place with right-left-right, cts 1-&-2.

IV. TRAVEL CW AND CCW; MAN SHOWS OFF

B 1-16 Repeat action of ms 1-16, part II with new partner and new corner person.

V. "HANDLESS CHAIN"

C 1-8 Repeat action of ms 1-8, part III.

VI. CIRCLE CW AND CCW

A 1-16 Repeat action of ms 1-16, part I.

Note: Phrasing of music may vary in melodies that accompany this dance. It may be necessary to begin a figure in middle of a phrase.

VARIATION

For those who would like an extra challenge, the following variation may be substituted for part I. This variation is more lively and will probably take some practice before it can be done smoothly.

Formation: 4 or more couples in circle A, facing center of circle with light handhold.

Steps: Change step, modified arm-swing turn,* and various steps for men.

* **Modified arm-swing turn**
See instructions for arm-swing turn. A modified arm-swing resembles usual arm-swing turn, but it is performed in circle A. On first step, turn to face your partner, swinging both arms to either side. Do not release handhold with partner or corner person:

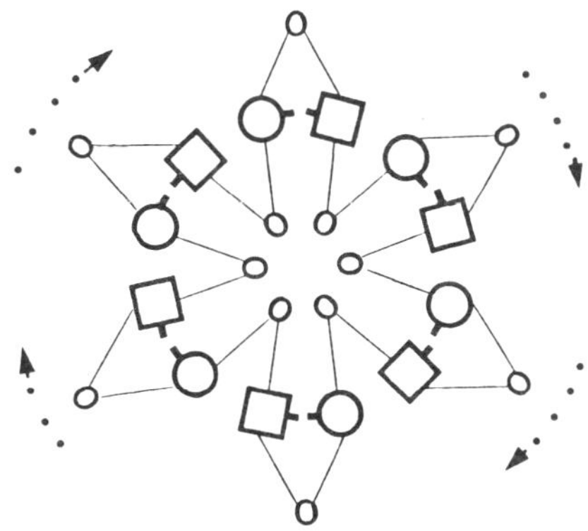

Perform first step pattern (step-close-step-pivot) such that man dances *backward* in CW LOD and woman *forward* in CW LOD. At end of pattern, turn away from your partner. Your arms should swing down and then out to either side. Do not release handhold with partner or corner person.

First arm-swing turn

You should now be facing your corner person. Perform second step pattern such that man dances *forward* in CW LOD and woman *backward* in CW LOD. At end of pattern, turn away from your corner person. Your arms should swing down and then out to either side.

You have just performed 2 modified arm-swing turns. You should now be facing partner.

MUSIC 2/4

Phrase	Ms	Part	
		I.	**CIRCLE CW AND CCW; MODIFIED ARM-SWING TURNS**
A	1-8		With man beginning left foot and woman right, perform 7 modified arm-swing turns in CW LOD. Following this, remain facing partner with arms extended to either side and stamp 3 times in place with right-left-right, cts 1-&-2.
	9-16		Everyone will now dance modified arm-swing turns in reverse direction. With man beginning left foot and woman right, perform 7 modified arm-swing turns in CCW LOD. Man begins by facing partner and dancing *forward.* Woman begins by facing partner and dancing *backward.* Following this, remain facing partner with arms extended to either side and stamp 3 times in place with right-left-right, cts 1-&-2.

* * *

Perform other parts of dance as before. Remember that part VI is same as part I.

Seierstad-hopsa

Namdalen, Nord-Trøndelag

Formation: 6 or more couples in 2 lines; men are on one side, women the other. Couples are numbered here for greater clarity. Man places left hand low on hip, fingers forward; right arm is relaxed by his side. Woman holds skirt to either side.

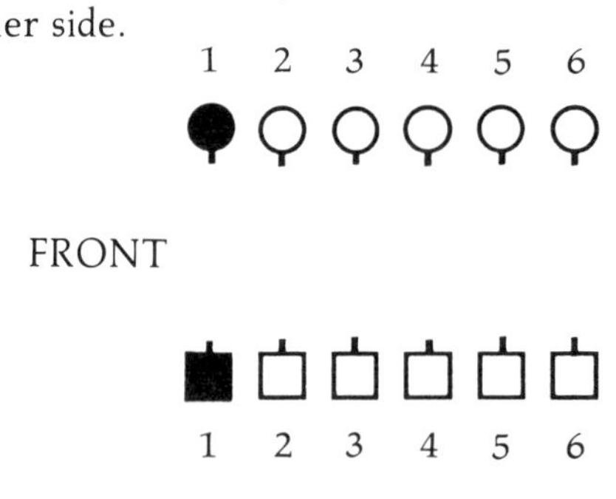

Steps: Change step, walking step

MUSIC 2/4 ______________________________

Phrase Ms Part

I. INVITATION TO DANCE; FORMAL GREETING

See instructions for formal greeting. No. 1 man will invite 2 women and 1 man to dance as follows:

Phrase	Ms	
A	1	Beginning left foot, no. 1 man takes 2 walking steps across to partner on each ct.
	2	No. 1 man closes left foot beside right. He bows briefly toward partner with formal greeting.
		No. 1 woman lowers with formal curtsey in 2 cts.
	3	On first ct, man pivots a little CW on right foot and crosses left foot in front of his right, moving toward no. 2 woman. On second ct, he then faces no. 2 woman, closing right foot beside left.
		No. 1 woman rises from curtsey in 2 cts.
	4	No. 1 man bows briefly toward no. 2 woman with formal greeting.
		No. 2 woman lowers with formal curtsey in 2 cts.

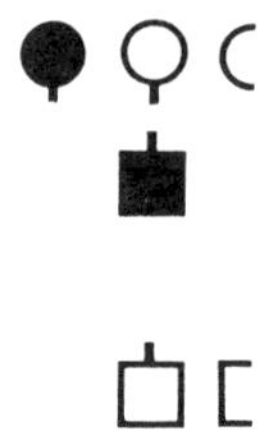

5 Pivoting quickly on right foot, no. 1 man makes ½ CCW turn and steps forward with left foot on ct. 1. He takes a second step toward no. 2 man on ct 2.

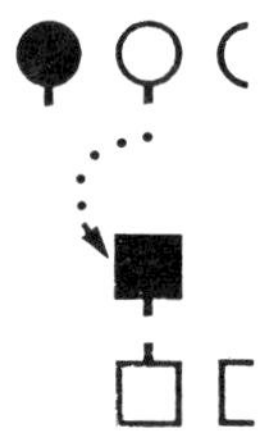

No. 2 woman rises from curtsey in 2 cts.

6 No. 1 man closes left foot beside right. He bows briefly toward man no. 2 with formal greeting.

No. 2 man bows forward in 2 cts.

7-8 Beginning left foot, no. 1 man takes 4 walking steps to his home position on each ct, making ½ CW turn as he does so. He should conclude in original starting position.

No. 2 man rises from bow in 2 cts.

Note: The above description reflects the way this part of the dance is currently performed in Norway. In her book, *Norske Folkedansar,* Klara Semb does not refer to any musical counts when explaining no. 1 man's actions.

II. DOWN BETWEEN LINES; RETURN

B 1 No. 1 and no. 2 couples will assume following handhold during first step on ct 1:

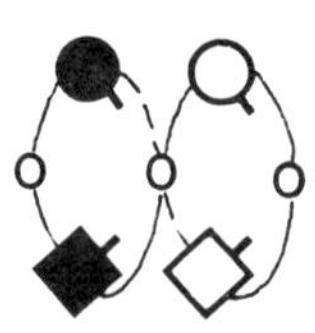

Note that no. 2 man's left hand and no. 1 woman's left hand are joined beneath no. 1 man's right hand and no. 2 woman's right hand. All hands are lifted about chest level.

Beginning left foot, perform 1 change step and begin to travel down between lines toward opposite end. Stamp on first step of pattern.

2-4 Beginning right foot, continue forward, dancing 3 change steps.

On last step, above dancers make ½ turn on right foot as follows:

Without releasing handhold, no. 1 man will lift up his right arm and pivot ½ CW turn beneath it. As he does this, he will exchange places with partner and conclude facing front of room.

Without releasing handhold, no. 1 woman will pass beneath uplifted arm of her partner (no. 1 man) and make ½ CW turn. She will conclude facing front of room on his left side.

No. 2 man makes ½ CCW turn in place. He should be to left of no. 1 woman and conclude facing front of room. No. 2 woman makes ½ CW turn in place. She should be to right of no. 1 man and conclude facing front of room.

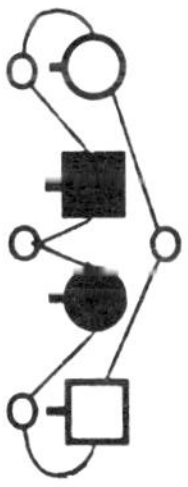

5-8 Beginning left foot, perform 4 change steps forward. On last step, no. 2 couple falls somewhat behind no. 1 couple in preparation for next part.

III. CHAIN

No. 1 and no. 2 couples release handhold. Partners turn to face each other.

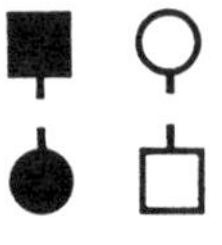

A 1-6 Take right hands with partner and chain past 6 persons, counting partner as no. 1. Beginning left foot, dance forward in tight circle, using 1 change step per person. (Men move in CCW LOD, women in CW LOD). Note that couples conclude in following position:

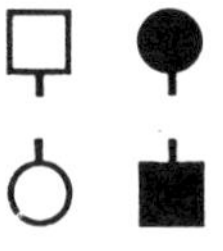

7-8 Partners meet and take right hands as in handshake. At same time, man places left hand firmly on woman's upper right arm.

While facing each other with above handhold, no. 1 couple quickly makes 1 CW turn in place with 2 change steps. Begin on left foot.

While facing each other with above handhold, no. 2 couple makes ½ CW turn in place with 1 change step. Partners then release handhold and move backward into their respective lines with 1 change step.

* * *

No. 1 couple has now progressed one space down set. They repeat part II with no. 3 couple and continue rest of dance as before. No. 3 couple performs actions of no. 2 couple as described above. No. 2 couple stands in place.

Each time dance repeats, no. 1 couple progresses 1 space down set. Soon this couple will have passed by 3 couples. When this happens, first 2 couples in line will then start the dance from beginning. No. 2 man invites his partner and neighboring couple to dance; he does this when no. 1 and no. 4 couples begin chain figure (part III). All active dancers will then begin part II simultaneously.

After no. 1 couple has danced with last couple, they retire into place at end of two lines.

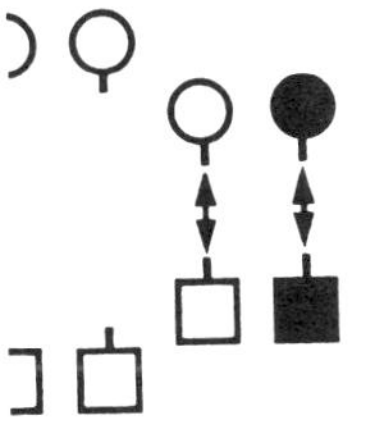

VARIATION

Once the figures are familiar to everyone, it is possible to start the dance with several groups of active couples. Several men (no. 1, 4, 7, 10, etc.) perform the formal greeting, part I. There will then be 2 or more groups of active couples dancing at the same time.

Seksmannsril

Six-person Reel
Asker, Akershus

Formation: 3 couples in circle A, facing center of circle with light handhold.

Steps: Step-hop

MUSIC 2/4

Phrase	Ms	Part
		I. CIRCLE CW AND CCW
A	1-8	Beginning left foot, perform 14 step-hops in CW LOD. Following this, stamp 3 times in place with left-right-left, cts 1-&-2.
	9-16	Beginning right foot, turn and perform 14 step-hops in CCW LOD. Following this, stamp 3 times in place with right-left-right, cts 1-&-2.
		II. EXTEND ARMS; CIRCLE CCW Release light handhold and turn on supporting right foot to face partner (man ¼ C turn, woman ¼ CCW turn).
B	1-8	On ct 1, clap hands together and quickly join hands with partner in position 6. Hands closest to center of circle are held low, while hands held away from center of circle are held high.

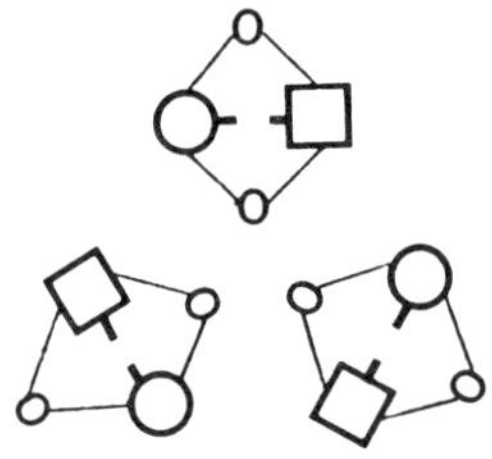

Beginning left foot on ct 1, perform 14 step-hops in CCW LOD with partner. Man dances forward, woman backward. Since man can see ahead, he helps to steer woman so that there is equal spacing between couples. Following this, stamp 3 times in place with left-right-left, cts 1-&-2.

Phrase	Ms	Part
		III. EXTEND ARMS; CIRCLE CCW Release handhold with partner and make ½ CCW turn on supporting left foot to face corner person.
	9-16	On ct 1, clap hands together and quickly join hands with corner person in position 6. Hands closest to center of circle are held low, while hands away from center of circle are held high.

Beginning right foot on ct 1, perform 14 step-hops in CCW LOD with corner person. Woman dances forward, man backward. Since woman can see ahead, she helps to steer man so that there is equal spacing between couples. Following this, stamp 3 times in place with right-left-right, cts 1-&-2.

IV. CHAIN

Release handhold with corner person and make ½ CCW turn on supporting right foot to face partner.

C 1-8 On ct 1, clap hands together and quickly take right hands with partner. Chain past 6 persons, counting partner as no. 1. Beginning left foot, perform 14 step-hops. Man moves in CCW LOD, woman in CW LOD. Stop in front of partner and stamp 3 times in place with left-right-left, cts 1-&-2.

V. CIRCLE CW AND CCW

Face center of circle and assume circle A with light handhold. Close right foot beside left with weight.

A 1-16 Repeat action of ms 1-16, part I.

VI. CLAP HANDS; CIRCLE CCW

Release light handhold and turn on supporting right foot to face partner (man ¼ CW turn, woman ¼ CCW turn).

B 1-8 Beginning with ct 1, clap hands together on each ct of music as if hitting cymbals together.

Beginning left foot on ct 1, perform 14 step-hops in CCW LOD with partner. Man dances forward, woman backward. Following this, stamp 3 times in place with left-right-left, cts 1-&-2. Do not clap on stamps.

VII. CLAP HANDS; CIRCLE CCW

Make ½ CCW turn on supporting left foot to face corner person.

9-16 Beginning with ct 1, repeat clapping action.

Beginning right foot on ct 1, perform 14 step-hops in CCW LOD with corner person. Woman dances forward, man backward. Following this, stamp 3 times in place with right-left-right, cts 1-&-2. Do not clap on stamps.

VIII. CHAIN

Make ½ CCW turn on supporting right foot to face partner.

C 1-8 Repeat action of ms 1-8, part IV.

Note: Phrasing of music may vary in melodies that accompany this dance. It may be necessary to begin a figure in middle of a phrase.

REPEAT DANCE FROM BEGINNING

Sekstur

Namdalen, Nord-Trøndelag

Formation: 6 or more couples in 2 lines; men are on one side, women the other. Couples are numbered here for greater clarity. Man places left hand low on hip, fingers forward; right arm is relaxed by his side. Woman holds skirt to either side.

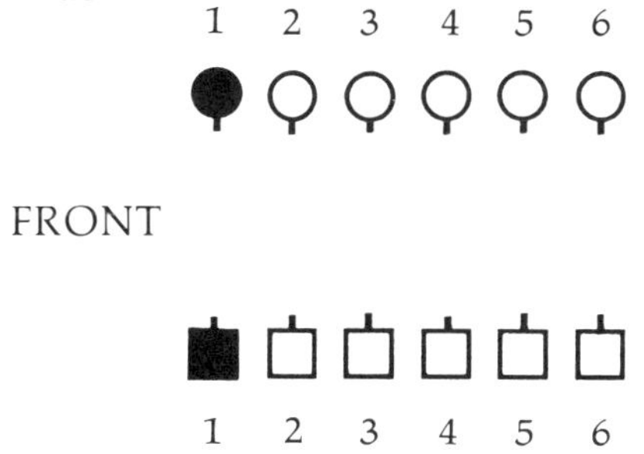

Steps: Touch step, walking step, waltz

MUSIC 3/4

Phrase	Ms	Part	
		I.	**INVITATION TO DANCE; FORMAL GREETING**
			See instructions for formal greeting. No. 1 man will invite 3 women and 2 men to dance as follows:
A	1-2		Beginning left foot, no. 1 man takes 2 walking steps across to partner on first ct of each ms.
	3-4		No. 1 man closes left foot beside right. He bows briefly toward partner with formal greeting. On first ct of next ms, he pivots a little CW on right foot and crosses left foot in front of his right, moving toward no. 2 woman.
			No. 1 woman lowers with formal curtsey in 6 cts.
	5-6		No. 1 man faces no. 2 woman, closing right foot beside left. He bows briefly toward her with formal greeting. On first ct of next ms, he pivots a little CW on right foot and crosses left foot in front of his right, moving toward no. 3 woman.
			No. 1 woman rises from curtsey in 6 cts. No. 2 woman lowers with formal curtsey in 6 cts.
	7-8		No. 1 man faces no. 3 woman, closing right foot beside left. He bows briefly toward her with formal greeting.
			No. 2 woman rises from curtsey in 6 cts. No. 3 woman lowers with formal curtsey in 6 cts.
	9-10		Pivoting quickly on right foot, no. 1 man makes ½ CCW turn and steps forward with left foot on ct. 1. He takes a second step toward no. 3 man on first ct of next ms.

No. 3 woman rises from curtsey in 6 cts.

11-12 No. 1 man closes left foot beside right. He bows briefly toward man no. 3 with formal greeting. On first ct of next ms, he pivots a little CW on right foot and crosses left foot in front of his right, moving toward no. 2 man.

No. 3 man bows forward in 3 cts.

13-14 No. 1 man faces no. 2 man, closing right foot beside left. He bows briefly toward man no. 2 with formal greeting.

No. 3 man rises from bow in 6 cts. No. 2 man bows forward in 6 cts.

15-16 Beginning right foot, no. 1 man takes 2 walking steps to his home position on first ct of each ms, making ½ CW turn as he does so. He then closes right foot beside left in home position on ct 3.

No. 2 man rises from bow in 6 cts.

Note: The above description reflects the way this part of the dance is currently performed in Norway. In her book, *Norske Folkedansar,* Klara Semb does not refer to any musical counts when explaining no. 1 man's actions.

II. CROSS TO OTHER SIDE; RETURN

No. 1, 2, and 3 men join hands with light handhold. No. 1 and no. 3 men place free hand low on hip, fingers forward.

No. 1, 2, and 3 women also join hands with light handhold. No. 1 and no. 3 women hold skirt to side with free hand.

These 2 groups will now change places with one another as follows:

B 1-7 Beginning left foot, men dance 7 touch steps along a curved path to opposite side. Footwork describes a half-circle in CW LOD.

Beginning left foot, women dance same figure as men, but their footwork describes half-circle in CCW LOD. Their line passes to inside of men's line.

FRONT

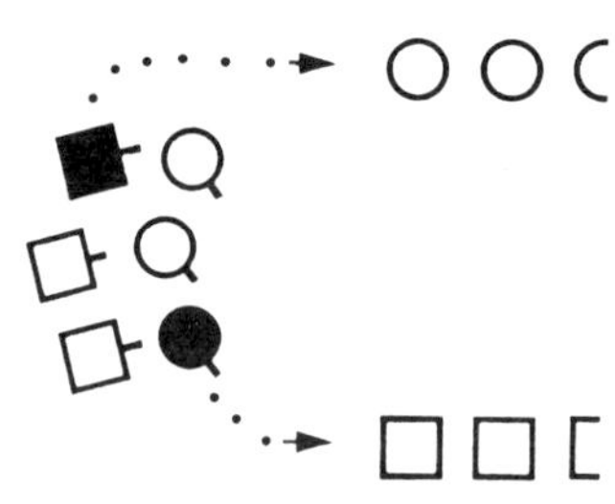

Note: Upon arriving at opposite side, the 2 lines should become straight. The third person in each line must turn a little and step backward on seventh step to accomplish this.

8 On ct 1, men and women in above lines step backward onto right foot, such that left foot remains in place and extends. Perform an acknowledgement.

9-16 Repeat action of ms 1-8, Part II, such that everyone returns to original positions.

III. CIRCLE CW AND CCW; REPEAT

Release handhold.

C 1-2 No. 1 and no. 3 couples quickly form a circle with light handhold. Beginning left foot, they dance 2 touch steps in CW LOD. (Dancers come into circle with first step.) On ct 3 of second step pattern, pivot on right foot and make ½ CW turn.

3-4 Beginning left foot, dance 2 touch steps in CCW LOD. On ct 3 of second step pattern, no. 3 man and woman release hands from circle.

5-6 No. 3 man and woman move backward into their respective lines with 2 touch steps.

No. 2 couple quickly joins no. 1 couple to form circle with light handhold. Beginning left foot, dance 2 touch steps in CW LOD. (No. 2 couple comes into ring with first step.) On ct 3 of second step pattern, pivot on right foot and make ½ CW turn.

7-8 Beginning left foot, dance 2 touch steps in CCW LOD.

IV. CHAIN

No. 1 and no. 2 couples release handhold. Partners turn to face each other.

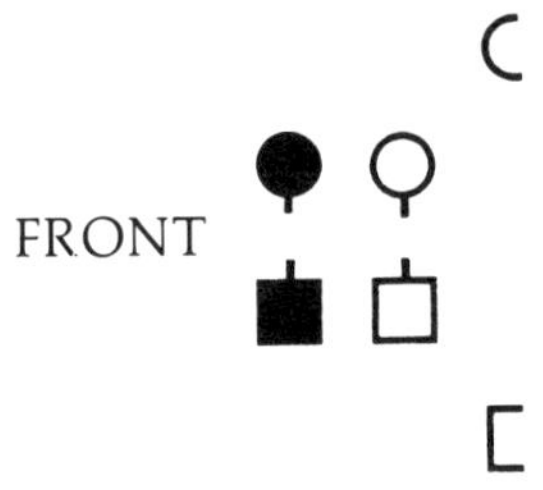

9-16 No. 1 couple assumes following handhold. Take right hands as in a handshake. At same time, man places his left hand on woman's upper arm or back.

Beginning left foot, no. 1 couple makes ½ CW turn with 2 touch steps, such that partners change places with each other. This action starts chain for no. 1 couple. They now continue to chain past 3 persons, dancing 2 touch steps per person.

No. 2 man and woman take right hands and chain past four persons, counting partner as no. 1. Beginning left foot, they dance 2 touch steps per person.

Conclude in following position:

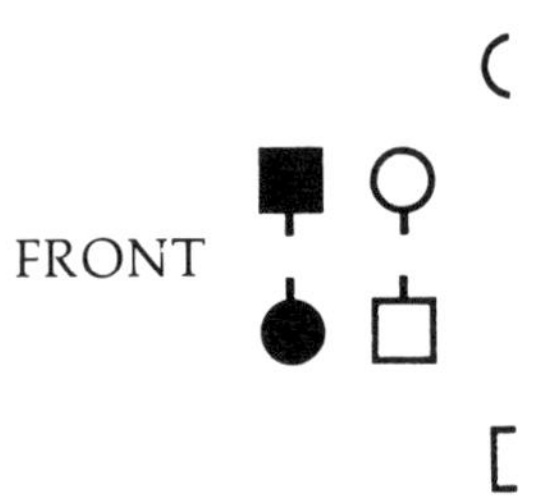

V. FORMAL GREETING

A 1-4 4 couples take part in this formal greeting. No. 1 and no. 2 couples will face partner and bow/curtsey. No. 3 and no. 4 men will turn toward each other and bow. No.3 and no. 4 women will turn toward each other and curtsey.

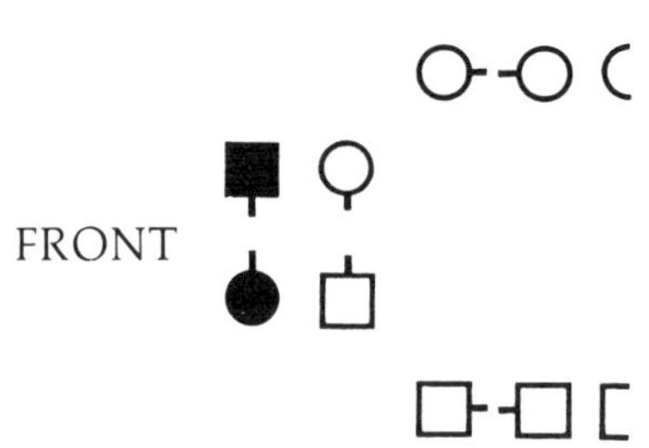

Men bow forward in 6 cts and rise in 6 cts.

Women lower in 6 cts and rise in 6 cts.

No. 3 and no. 4 persons turn back to original positions.

VI. WALTZ

5-14 No. 1 and no. 2 couples: partners come together with 1 waltz step to assume position 1. Man begins right foot and dances toward woman; woman begins left foot and dances mostly in place. Couples then waltz for next 9 ms, traveling around their own small circle a total of 1½ times. Note final positions:

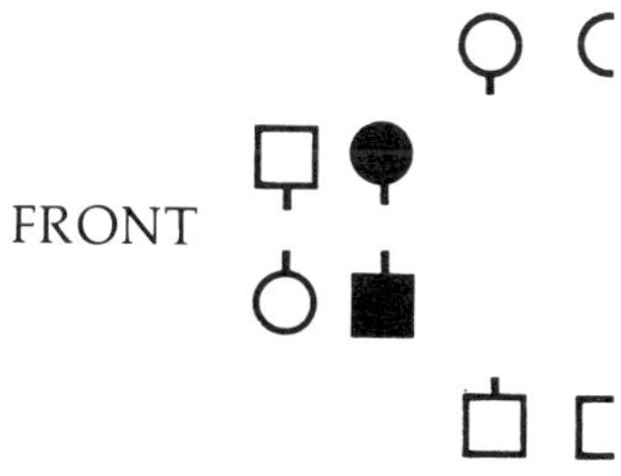

15-16 Partners release position and dance 2 touch steps backward into their respective lines.

* * *

No. 1 couple has now progressed one space down set. They repeat part II with no. 3 and no. 4 couples and continue rest of dance as before. No. 3 couple performs actions of no. 2 couple, as described above, and no. 4 couple performs actions of no. 3. No. 2 couple stands in place.

Each time dance repeats, no. 1 couple progresses 1 space down set. Eventually this couple will have passed by 3 couples. When this happens, these 3 couples will then start dance from beginning. No. 2 man invites his partner and neighboring couples to dance; he does this when no. 1 and no. 5 couples begin their greeting (part V) before waltz. All active dancers will then begin part II simultaneously.

When no. 1 couple reaches last couple at end of the 2 lines, parts II and III will be done by only 2 couples instead of 3. In part III, these 2 couples will dance *four* touch steps in CW LOD and then *four* touch steps in CCW LOD.

After waltz, no. 1 couple retires into place at far end of set.

VARIATION

Once the figures are familiar to everyone, it is possible to start the dance with several groups of active couples. Several men (no. 1, 5, 9, 13, etc.,) perform the formal greeting, part I. There will then be 2 or more groups of active couples dancing at the same time.

Part II. Acknowledgement

Part V. Formal greeting

Tantoli

Austfold

Formation: Several couples in circle B, facing CCW LOD with following position:

Steps: Change step, reinlender step, step-hop

MUSIC 2/4

Phrase	Ms	Part	
		I.	**HEEL-TOE SWING; CHANGE STEP** Man places weight on right foot, woman left foot.
A	1		On ct 1, swing free leg forward (man's left, woman's right), such that heel touches floor with toes pointing up. Upper body tilts a little backward as leg swings forward. Free leg is straight; supporting leg is bent somewhat.
			On ct 2, swing free leg backward (man's left, woman's right), such that toes touch floor with heel pointing up. Upper body tilts a little forward as leg swings backward. Free leg is straight; supporting leg is bent somewhat.
	2		With man beginning left foot and woman right, perform 1 change step forward.
	3-4		Repeat action of ms 1-2 with opposite footwork.
	5-8		Repeat action of ms 1-4.
		II.	**POSITION 2; INTRODUCTORY TURN**
B	1		Beginning left foot, man performs 1 reinlender step almost in place, turning a little CCW as he does so. At same time, he draws his partner forward and in front of him to assume position 2.
			Beginning right foot, woman dances in front of man and turns to face him with 1 reinlender step. By end of ms, partners should be in position 2. Man should be approximately facing CCW LOD.

2 Couple makes about ¼ CW turn with 1 reinlender step. Man should conclude facing away from center of circle, woman toward.

3-4 With man beginning left foot and woman right, make 1 CW turn with 2 reinlender steps.

III. COUPLE TURNS CW

5-7 With man beginning left foot and woman right, make about 3½ CW turns with 6 smooth step-hops. Extra half turn means that couple must rotate more quickly than usual on last 2 step-hops. Man should conclude facing toward center of circle, woman away.

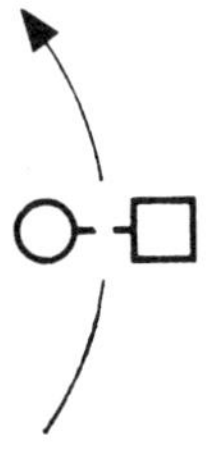

IV. LIFT

Man securely places hands on either side of woman's waist. Woman places both hands firmly on top of man's shoulders. Man and woman place weight on both feet.

8 On ct 1, man lifts woman by waist, using good flexion in knees, and makes ½ CW turn with her in air. On ct 2, he puts woman down on other side.

Woman prepares for lift by bending knees a little and then by jumping vertically on ct 1. As woman jumps, she also presses down on man's shoulders, partially supporting her weight. Woman lands with flexion in ankles and knees.

Note: This lift is sometimes omitted by dancers in Norway. They perform 8 step-hops instead of 6 and make 4 CW turns.

REPEAT DANCE FROM BEGINNING

Totur

Holt, Aust Agder

Formation: 2 couples in circle A, facing center of circle with light handhold.

Steps: Gesture step, step-behind

MUSIC 2/4

Phrase	Ms	Part
		I. CIRCLE CW AND CCW
A	1-3	Beginning left foot, perform 6 step-behind patterns in CW LOD. Everyone should lean a little outward from center of circle with counterbalance.
	4	Stamp 3 times in place with left-right-left, cts 1-&-2.
	5-8	Repeat action of ms 1-4 in opposite direction with opposite footwork.
		II. REFRAIN: GESTURE STEP
		Release light handhold and place hands on hips, fingers forward. Partners face each other and form a square. Place weight on right foot. Above happens during extended introductory note in music.

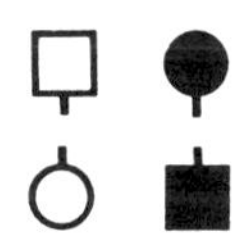

Phrase	Ms	Part
B	1	Beginning left foot, perform 1 gesture step pattern to right.
	2	Beginning right foot, perform 1 gesture step pattern to left.
	3	Beginning left foot, perform 1 gesture step pattern to right.
	4	Beginning right foot, perform 1 gesture step pattern to left, making following change:

Man begins to turn CW on third step of pattern and then hops on supporting right foot, pivoting 3/4 CW turn. He should conclude facing corner person.

Woman begins to turn CCW on third step of pattern and then hops on supporting right foot, pivoting 3/4 CCW turn. She should conclude facing corner person.

direction of turns 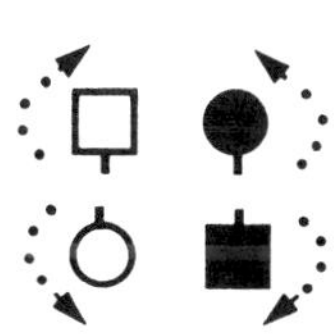 TO 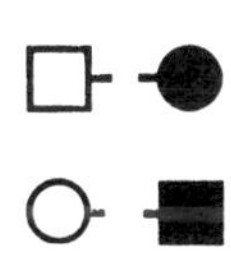

It may be helpful to remember that partners turn outward and away from each other.

5-8 Facing corner person, repeat action of ms 1-2, part II, two times through.

III. ONE-HAND CROSS

Face center of circle. Men extend right arms and interlock right hands by grasping around base of thumb area and onto upper surface of hand. Women do same, joining hands below men's handhold. Place left hand on hip, fingers forward.

A 1-4 Repeat footwork of ms 1-4, part I. Everyone should lean a little outward from center with counterbalance.

5-8 Switch hands (left hands joined, right hand on hip). Repeat footwork of ms 5-8, part I.

IV. REFRAIN: GESTURE STEP

B 8 Repeat action of ms 1-8, part II.

V. TWO-HAND CROSS

Face center of circle. Men extend both arms and join hands by hooking their fingers together, one hand above, one below. Women do same, joining hands below men's handhold. To avoid confusion, each person should turn left palm up and right palm down when taking hands. Everyone should lean a little outward from center of circle with counterbalance.

A 1-4 Repeat footwork of ms 1-4, part I.

5-8 With same handhold, repeat footwork of ms 5-8, part I.

VI. REFRAIN: GESTURE STEP

B 1-8 Repeat action of ms 1-8, part II.

VII. CLAPPING

A 1-4 Face center of circle. Beginning with ct 1, clap hands together on each ct of music as follows: hold left hand in front of chest (palm facing up) and clap right hand onto it. Repeat footwork of ms 1-4, part I.

5-8 Continue clapping in same manner. Repeat footwork of ms 5-8, part I.

VIII. REFRAIN: GESTURE STEP

B 1-8 Repeat action of ms 1-8, part II.

IX. CIRCLE CW AND CCW

Face center of circle and join hands with light handhold.

A 1-8 Repeat action of ms 1-8, part I. Following last stamp, perform an acknowledgement toward center of circle.

Tretur

Fana, Hordaland

Formation: 12 persons in square. On each side, 1 man stands between 2 women.*

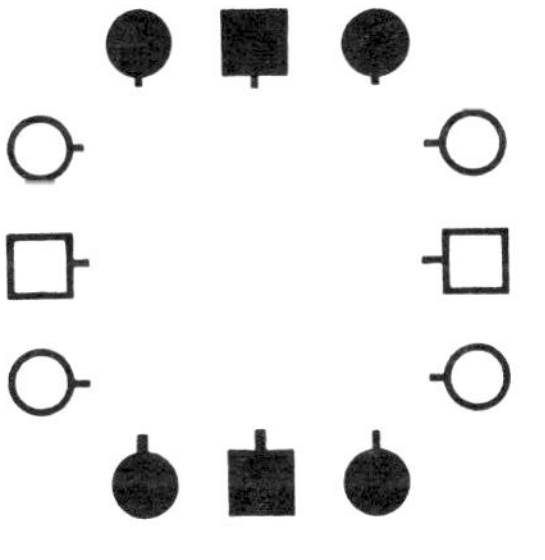

Women join inside hands behind man's back. Man takes outside hands of women with light handhold (his left hand to woman on left, right hand to woman on right). All hands are approximately shoulder high.

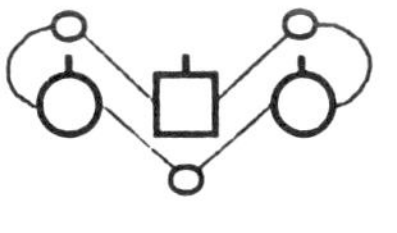

*May also be performed with 1 woman between 2 men.

Steps: Walking step.

MUSIC 2/4

Phrase Ms Part

I. FORWARD AND BACK

A 1-4 Beginning left foot, head trios (those facing front of room and those with their backs to front) go forward with 4 walking steps on cts 1-2-1-2. Acknowledge dancers opposite.

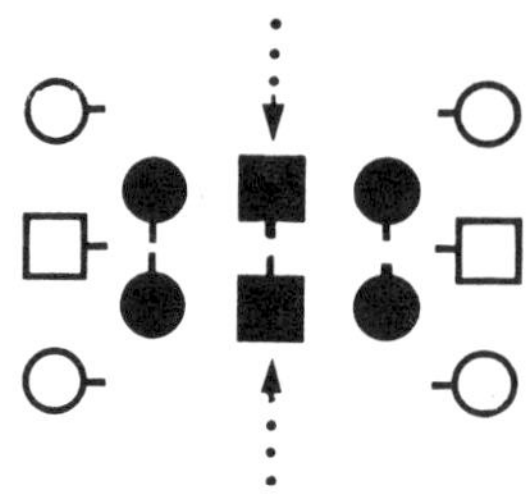

		Take 4 walking steps backward to places on cts 1-2-1-2. During part I, man flirts with both women.
	5-8	Side trios repeat action of ms 1-4 from their respective positions.
	9-16	Repeat action of ms 1-8, part I.

II. ARMS LIFT; TURNS

Note: This part is best described as an 8 ct sequence. (4 ms of 2/4 time). The directions are written accordingly.

B	1	Women lift arms as a gate for man to pass beneath. Man stamps with left foot on ct 1 and bends forward a little on ct 2. He goes backward through gate on right foot.

Middle person goes backward beneath arms. (As shown, woman is middle person on right side of photo and man is middle person on left side of photo.)

	2	Everyone lifts hands well above head level. Man curves around left-hand woman, taking 2 steps on cts 3-4 and leading with his back side. As man curves around, left-hand woman is lightly pulled under uplifted arms and makes 1 CW turn with several steps in place. Man will end up partially in front of left-hand woman.

At same time, right-hand woman begins to turn in CCW direction.

3	Man moves in front of the women with 2 steps on cts 5-6. As he comes to that position, right-hand woman completes her CCW turn with several steps in place. Arms are still raised.
4	Man makes 1 CW turn beneath uplifted arms with 2 steps on cts 7-8. Trio should conclude in original position, lowering hands to shoulder level.
	Man flirts with women during above figure and they with him.
5-16	Repeat action of ms 1-4, part II, three times.
	Note: Dancers do not release hands at any time.

REPEAT DANCE FROM BEGINNING

Tretur

Hordaland

Formation: 1 man and 2 women in circle, facing center of circle with light handhold.*

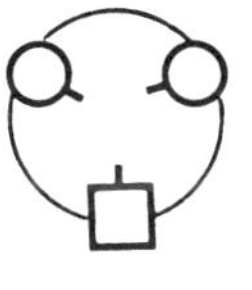

Steps: Extending steps, reel step, step-hop, walking step.

* May also be done by 2 men and 1 woman.

MUSIC 2/4

Phrase	Ms	Part
		I. CIRCLE CW AND CCW
A	1-8	Beginning left foot, perform 14 step-hops in CW LOD. Following this, stamp 3 times in place with left-right-left, cts 1-&-2.
	9-16	Beginning right foot, turn and perform 14 step-hops in CCW LOD. Following this, stamp 3 times in place with right-left-right, cts 1-&-2.
		II. FORM LINE; FLIRTATION Release handhold. Man folds arms above chest and parallel to ground. Woman holds skirt to either side or places hands on hips, thumbs forward.
B	1-8	Beginning left foot, man may choose to dance 14 extending steps or 14 reel steps. With first step or so, man comes between the 2 women, facing woman

who was on his right in ring. Man concludes with 3 stamps in place, left-right-left, cts 1-&-2.

second woman  first woman

Beginning left foot, *first* woman will also perform 14 extending steps or 14 reel steps, but she does the step the man has *not* chosen. (Thus if man does extending step, she dances reel steps and vice versa.) She concludes with 3 stamps in place, left-right-left, cts 1-&-2.

Beginning left foot, *second* woman makes ¼ CCW turn and takes 4 walking steps forward.

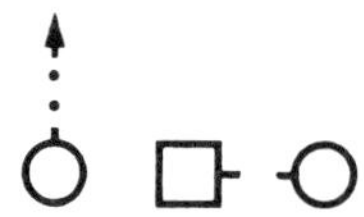

On last step, she makes ½ CW turn, pivoting on right foot.

Beginning left foot, second woman takes 4 walking steps in other direction. On last step, she makes ½ CCW turn, pivoting on right foot.

Second woman repeats above actions, walking back and forth.

Note: During above figure, both women try to win attention of man. First woman smiles at man. Second woman acts in a jealous manner, perhaps making faces at first woman.

9-16 Man repeats action of ms 1-8, part II, now facing *second* woman. With first step or so, he makes ½ CW turn to face that woman. He remains facing her after the 3 stamps.

Women reverse roles and repeat action of ms 1-8, part II. Remember that second woman does step man has *not* chosen.

III. FIGURE EIGHT

Footwork for this figure should describe a figure eight pattern on floor. Man goes to right and loops around one end of figure eight. Woman facing him goes to right and crosses center of figure eight. Woman behind man goes to left and will also cross center of figure eight, but she waits for other woman to go by first. It is helpful if she waits about 3 cts before starting.

1-16 Beginning left foot, perform 2 figure eights with 30 quick walking steps. Dancers should conclude figure eight in starting positions. They then stamp in place with left-right-left, cts 1-&-2.

Note: Women do not hesitate to show their dislike for each other during above figure. When passing by man, however, each of them smiles.

REPEAT DANCE FROM BEGINNING

Assume original ring formation and repeat dance from beginning. During part II, however, woman who was to right side of man will now be middle person. She faces man first, deciding which step to start with. Man dances other step.

Repeat dance again. This time other woman will be middle person during part II. Conclude *tretur* by performing only ms 1-8, part I.

Åttetur

Asker, Akershus

Formation: 4 or more couples in circle A, facing center of circle with light handhold.

Steps: Lilting step, waltz

MUSIC 3/4

Phrase Ms Part

I. CIRCLE CW AND CCW

A 1-8 Beginning left foot, dance 8 lilting steps in CW LOD. On last step, pivot on supporting right foot and make ½ CW turn; left foot remains lifted during turn.

9-16 Beginning left foot, dance 8 lilting steps in CCW LOD.

II. CHAIN; REVERSE

Turn to face partner.

B 1-7 Take right hands with partner and chain past 4 persons, counting partner as no. 1. Beginning left foot, dance 1 lilting step with partner and then 2 lilting steps with each person after. Men move in CCW LOD, women CW LOD.

8-9 Take right hands with person number 5 and make ½ CW turn with that person, such that you change places. This is done with 2 lilting steps. During turn, man places left hand low on hip, finigers forward. Woman holds skirt to side with right hand.

10-16 Take left hands with next person (beyond the one you turned with) and continue the chain, now traveling back the way you came, past 3 persons. Beginning right foot, dance 2 lilting steps per person. Men move in CW LOD, women in CCW LOD.

On last ms, stop in front of partner with one step on right foot. Close left foot beside right on ct 3.

III. INFORMAL GREETING

See instructions for informal greeting.

C 1-2 Man bows forward in 3 cts and rises in 3 cts.

Woman lowers with informal curtsey in 3 cts and rises in 3 cts.

IV. WALTZ; ACKNOWLEDGEMENT

3-14 Partners come together with 1 waltz step to assume position 2. Man begins right foot and dances toward partner; woman begins left foot and dances mostly in place. Couples then waltz for the remaining 11 ms, making about 5½ CW turns in overall CCW LOD.

15 Partners now dance 1 waltz step almost in place as they turn into their original starting positions in ring A. While this happens, they open away from each other. Woman slides left hand along man's right arm to his hand. Man holds arm out to steady her.

16 Woman holds skirt to side with right hand. She performs an acknowledgement toward partner, placing right foot behind left.

Man places left hand low on hip, fingers forward. He performs an acknowledgement toward partner, placing left foot behind right.

Old-Time Couple Dances

An assortment of dances thrived in Europe and North America during the 19th century. Besides country-dances, quadrilles, and other set dances, a new generation of couple dances came into vogue at this time. Most celebrated of these were the Viennese waltz and the polka. All of the 19th century dances, both set dances and couple dances, were eventually known as "old-time dances" in Great Britain and the United States.[1]

As mentioned in the previous chapter, set dances from the 19th century were called *turdanser* or "figure-dances" in Norway. Couple dances from that period, on the other hand, were called *gammeldanser.* Literally meaning "old dances," *gammeldanser* may be translated as old-time dances—or more specifically, old-time couple dances.

Face to Face

The most popular old-time couple dances in Norway were the Viennese waltz, the polka, the Parisian polka, the polkamasurka, the masurka, the springpolka, and the reinlender. All of these had one important feature in common. This was a figure in which partners held onto each other, face to face, with a certain dance position—such as that found in the Viennese waltz—and then whirled around together, all the while progressing around the floor with everyone else.

Old-time couple dances were not delivered on Norway's doorstep all at once. They came calling one at a time throughout the 19th century, each being welcomed in its turn. Most came from other countries. Over the years, these dances were given a certain Norwegian style. A few were "Norwegianized" almost beyond recognition.

The Waltz: "An Outright Romp"

The Viennese waltz was the first to arrive in Norway. Exactly where the dance originated is uncertain, but it is thought to have been inspired by traditional couple dances from the alpine regions of Germany and Austria. Some of these couple dances, such as the *ländler* of Upper Austria, have similar turning figures.[2] The name *waltz* was apparently derived from the old high German *walzan,* which meant to turn or roll.[3]

The earliest reports of the Viennese waltz describe it as being one of the most fashionable dances in Prague and Vienna during the 1790s. It has been associated with the second of those cities ever since.[4] From Prague and Vienna the dance soon spread to the rest of Europe, including Norway. After attending a ball in Kristiania, one Englishman, Thomas R. Matthus, wrote in his diary of 1799:

> *Otter and myself, finding that we made but a poor figure at the Walse, did not dance much.*[5]

Matthus could be excused if his knowledge of the waltz was less than perfect. At the time he wrote, the dance was still a novelty and not everyone approved of it. Conservatives among the upper classes, in fact, attacked it from all sides. Some claimed that waltzing was unhealthy. In a German pamphlet from 1799, Salomo J. Wolf wrote that "too much dancing, especially waltzing, is as injurious to the soundness of the mind as to the health of the body. . . ."[6] George Yates objected to the form of the dance in his book, *The Ball: or a Glance at Almack's* (1829):

> *It is nothing in short but an outright romp, as destitute of figure or variety as the motion of a horse in a mill.*[7]

More than anything else, the upper classes criticized the Viennese waltz for the way partners held onto each other. It must be remembered that they were accustomed to dances in which partners kept a discreet distance from each other, perhaps joining hands now and then. In their minds, the basic position of the Viennese waltz was simply indecent.[8]

Some of the upper classes, especially in France and England, preferred a very different waltz, sometimes called the French waltz, which they thought to be less risqué. It consisted mostly of figures in which partners joined hands and turned beneath their uplifted arms in a variety of configurations.[9] Like the Viennese waltz, this waltz appears to have been inspired by Alpine dances from Germany or Austria.[10]

The two waltzes became rivals, but there was really no contest. The Viennese waltz proved to be irresistible. It could be seen in many of Europe's finest ballrooms by the middle of the 19th century.[11]

The upper classes in Norway went so far as to include the Viennese waltz as a figure in some of the dignified country-dances from England. This idea does not seem to have occurred to the English. Although they used waltz melodies to accompany some of their country-dances, which were subsequently called "waltz country-dances," the English did not include the dance itself.[12] It was probably with some interest, therefore, that one Englishman, Edward D. Clarke, witnessed a different way of performing country-dances at balls in Kristiania. In his travelogue of 1799, he wrote:

There are public balls on a Sunday evening, once in every fortnight. These are held in a large room belonging to the principal inn; and the ball is followed by a supper. Tickets are given to the different persons as they enter, to regulate their places in the dance; a different set of tickets being distributed for a similar purpose at supper. The dances are, the waltz, which always has the preference, and the common English country-dance; but even in the country-dance the waltz is introduced: it is so great a favourite, that our English dance would probably not be tolerated, but in compliment to the English who may happen to be present.[13]

One of the better-known examples of a country-dance with waltzing in Norway is the *sekstur* from Namdal.

Out in the countryside of Norway, people took to the Viennese waltz like ducks to water. Many of them were already acquainted with other dances, called *bygdedanser* or "provincial dances," in which partners turned together. Consequently, they did not find the Viennese waltz as difficult to accept as did some of the upper classes. The dance soared in popularity. Norwegian fiddlers responded with dozens of new waltz melodies.

Today the Viennese waltz is sometimes confused with another waltz that developed in the United States during the early 1900s. Often called the "box-step waltz" (because its steps trace a square pattern on the floor), it is slower than the Viennese waltz and is done almost in place with little turning.[14]

Polka Delirium

Toward the middle of the 19th century, other couple dances followed in the wake of the Viennese waltz, but none would compare with the polka. The dance caused such a sensation in European society that contemporary sources spoke of the "polka delirium" and the "polka mania."[15] Probably the greatest attraction of the polka was a turning figure in which partners almost galloped around the floor at a brisk pace. The dance is best remembered for this figure, but at one time it had other figures as well.[16] For instance, agile couples might stop turning and do a heel-toe gesture with either foot.[17] All of the figures were performed in a free-style manner. This was clearly indicated in some instructions for the polka that were printed in an 1844 issue of the *Illustrated London News:*

> *La Polka, like its predecessors the waltz and galop, is a* danse à deux, *couples following each other in the* salle de dansé, *commencing at pleasure and adopting, of the following figures, that which pleases them most at the moment.*[18]

Partners, of course, were obliged to do the same figure at the same time. To avoid confusion, it was usually the man who decided what figure would be danced and when. This arrangement was also noted in the *Illustrated London News:* "Change of figures left to gentleman."

The melodies that accompanied the polka were typically light and spirited, adding a touch of excitement to the dance. Unlike waltz melodies, which had a 3/4 rhythm, polka melodies had a 2/4 rhythm.

The polka is believed to stem from a genre of couple dances that were done by the common people in Eastern Bohemia, now a province of Czechoslovakia, during the early 19th century. It came to Prague around 1830. The earliest known description of the dance was written in Prague by Jaroslav Langer for a scholarly journal in 1835. Four years later, the polka arrived in Vienna, where it became an overnight success.[19]

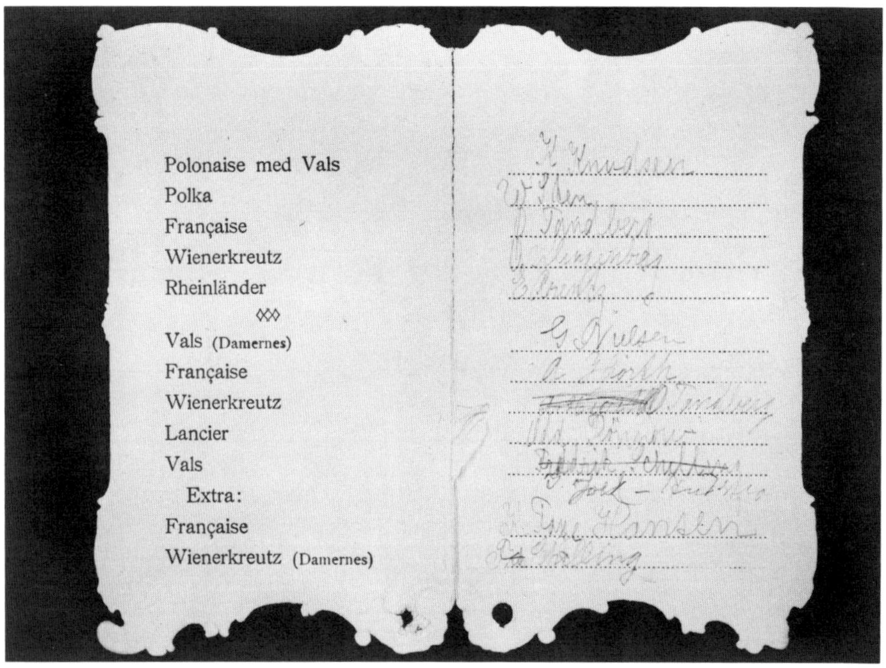

A program for a ball held in Kristiania, 1896. At such a ball, men attempted to find a partner for each dance in advance. Their names were then written down on the woman's program, as shown at right. Handskriftsamlingen, Universitetsbiblioteket, Oslo.

The polka did not escape the notice of the upper classes in Norway. They made every attempt to keep up with the latest trends from abroad. Precisely when the polka came to Norway is not known, but there are scattered references to it from the second half of the 19th century. Among these are program notes for formal balls, where the polka was a regular guest.[20]

To what extent the polka was done in the countryside of Norway is uncertain. Accounts are quite scarce. The earliest is found in a travelogue, *A Long Vacation Ramble in Norway and Sweden,* written by John Willis Clark and J. W. Dunning in 1856. The authors visited a traditional wedding in Rusaanes, in Northern Norway, where they saw "many waltzes and polkas."[21]

It should be mentioned that the Norwegians did have several regional folk dances that resembled the polka in both form and rhythm, but these were not traditionally known as polkas. Two examples are the *hamborgar* and *hoppvals*. They differed from the polka mostly in regard to stylistic features. All of the polka-like dances apparently came into existence during the 19th century, but little else is known of their background. Perhaps it was one of these dances that Clarke and Dunning saw in northern Norway.

Three Unusual Polkas

Oddly enough, the Norwegians affixed the name *polka* to three other dances that had little in common with it. They were the *springpolka,* the *Parisian polka,* and the *polkamasurka.*

The springpolka was an offshoot of the *hambo,* an old-time dance from Sweden. It appears to have crossed the border into Norway's eastern provinces around the turn of the 20th century. Some Norwegians, especially those who lived close to the Swedish border, continued to call the dance by its Swedish name.[22] Others called it the springpolka, a name that was evidently coined in Norway. *Springpolka* roughly translates as "the springing polka."

Partners began the springpolka by standing to the side of each other and joining one or two hands. They then performed two step-swings, took a few steps forward, and turned together several times in a closed position. Springpolka melodies were quite distinct from usual polka melodies in that they had three beats per measure instead of two.

Two girls from the Hardanger province, photographed in 1912. The accordion (and several instruments of similar construction) developed in Europe during the 19th century and often accompanied old-time couple dances.
Norsk Folkemuseum.

A Norwegian old-time band.

The Parisian polka, or *parisarpolka,* as the Norwegians call it, is rather puzzling. While its name would suggest the dance was imported from Paris, there is no record of it in French sources, including Desrat's comprehensive *Dicionnaire de la danse,* published in 1895.[23] It seems possible that the dance may have been done in several countries at one time, but was preserved only in Norway. This thought is given some support by a statement in Albert Zorn's *Grammatik der Tanzkunst,* published in Leipzig in 1887: "The Tour or Parisian Polka, once so popular, has long since disappeared. . ."[24] Unfortunately, Zorn said nothing else about this Parisian polka, making it difficult to know whether it was related in any way to Norway's Parisian polka.

Regardless of where it started, the Parisian polka found a home in the fjord country of western Norway, as well as in two northern provinces, Trøndelag and Nord-Norge. There were two basic figures. During the first, the partners joined one or two hands and danced a few steps forward. They then turned separately and danced the same number of steps in the opposite direction. During the second, partners briefly turned together. These two figures were repeated in the same order for as long as the music lasted. Parisian polka melodies —as with polka melodies elsewhere— had a 2/4 rhythm.

The Parisian polka sailed to North America with Norwegian immigrants around the turn of the 20th century. There it became one of the more popular dances among Scandinavians who settled in the midwestern and northwestern states. Today it is often called the "Scandinavian polka" or "Seattle polka."[25]

Yet another polka, the *polka-mazurka,* was adopted by Norwegians. This dance was first popularized in Paris about 1845.[26] Whether it was invented in that city is no longer known. Following its debut in Paris, the polka-mazurka embarked for other destinations. In his book, *Norske Dansetradisjonar,* Egil Bakka gives some idea of when it was done in Norway:

> *While the polka-mazurka was much used in the dance schools, it appears to have quickly won a place for itself in rural communities as well: Johan Christiansen, born in 1832, reports that he learned several dances, including the polka-mazurka, from the neighbor girls as a young boy, which can be hardly later than 1850. We have another source from 1851.*[27]

The polka-mazurka was one of several European dances (such as the galop-polka and the mazurka-waltz) that combined parts of two earlier dances, thus making a third. In this case, a basic step from a Polish dance, the mazurka,[28] was combined with the turning figure of the polka. Both were done to mazurka melodies, which had a 3/4 rhythm. Partners did the polka-mazurka face-to-face with the waltz position or a similar position. They traveled sideways with several mazurka steps and then made a half turn with a modified polka step, repeating this pattern throughout the dance.[29] In Norway, partners could make half or whole turns.

Not to be confused with the Polish mazurka was the Norwegian *masurka,* sometimes called the *austlandsk* or "eastern" *masurka.* This dance appears to have been nothing more than a more energetic version of the polka-mazurka. The basic form of the dance was the same—partners traveled sideways and then made half or whole turns—but liveler steps were used in place of the mazurka and polka steps. The Norwegian masurka was traditionally found in Valdres and other east-central provinces. The dance probably developed from the polka-mazurka sometime around the turn of the 20th century.[30]

The Reinlender: "One from the Rhineland"

No discussion of Norwegian old-time dances would be complete without some mention of the reinlender. This zesty dance has become a real favorite with Norwegians, who perform it from one end of their country to the other.

The name *reinlender* may be translated as "one from the Rhineland." The earliest accounts of the dance are found in Germany, where it was given several names. In his book, *Grammatik der tanzkunst,* Zorn wrote:

> *In the year 1850 there appeared, in all parts of Europe, the "Schottische," a round dance which had, as early as 1844, been executed under the name "Rheinländer," and in the Rhenish*

countries it was known as the "Bavarian Polka." These names in some places have been retained until this time. The music, which is in 2/4 measure, is rendered very slowly, with the effect of a 4/4 time.[31]

Zorn's statement leaves us wondering why people in Bavaria associated the dance with lands along the Rhine river, while people who lived in those lands associated the dance with Bavaria. Even more perplexing is why others associated the dance with Scotland. *Schottische* is the German word for Scottish. Since the dance was almost certainly a German creation, "essentially German in character and music," as pointed out by Routledge's *Ballroom Guide* from about 1860,[32] the reason for calling it Scottish is now unclear. Moreover, the dance did not arrive in Scotland until some years after it was known in Germany. Yet *schottische* remains the most common name for the dance in Europe and North America today.

There were two versions of the schottische. One of these is found in several dancing manuals from the second half of the 19th century, such as Zorn's work, which was published in 1887. The manuals describe a schottische with two figures, both of which were done with the waltz position. Partners began the first figure by traveling forward with three steps (step-close-step-hop). The same pattern was then repeated in the opposite direction. For the second figure, partners made two clockwise turns with four step-hops.

A second version of the schottische appeared during the late 19th century. Often called the "barn dance" or "military schottische,"[33] it is believed to have originated in the United States.[34] This schottische was similar to the other, but the first figure was done somewhat differently. Instead of assuming the waltz position, partners stood side-by-side, facing forward. The man placed his right arm around

A Norwegian reinlender with variations.

the woman's waist, and she rested her left hand on the man's nearest shoulder. They then traveled forward—not to the side—with two of the basic steps mentioned above.

The second schottische was quickly taken up in several parts of Europe and is still quite popular today. It is especially widespread in Scandinavia, where it has blossomed into numerous regional variants. The dance is known as the *skottis* in Iceland, the *schottis* in Sweden, the *sotiisi* or *jenkka* in Finland, the *rheinländer* in Denmark, and the *reinlender* in Norway—not to mention the many dialectical names that are found in these countries.

Dance and Holiday Traditions

Old-time dances became a part of traditional Norwegian society. In some areas the dances were so well-received that they nudged aside older provincial dances. Near the end of the 19th century, it became increasingly common to see old-time couple dances at social gatherings of every sort, from holiday events to weekend meetings with friends.

Two holidays that usually called for some dancing were Christmas and St. Hans evening—better known to us as Midsummer's Eve.

There was generally no dancing during the few weeks before Christmas. Everyone was busy scrubbing floors, chopping extra firewood, baking Christmas foods, and making other preparations for the upcoming holiday. Following Christmas day, everyone could relax and have a good time for a week or so. They visited family and friends, keeping themselves entertained by dancing and playing games. Ample supplies of food and drink were always on hand. Part of the Christmas festivities included the tradition of the *julebukk* or "Christmas goat." Young people would disguise themselves and go from house to house in search of treats—not unlike our Halloween tradition. Anne Odegård wrote that there could be dancing as well:

> *For julebukk the people would go out in improvised costumes, most often old clothes with an old leather cap pulled well down over the eyes, and go visiting houses. Some places they would get refreshments, and if there was someone who could play the fiddle or harmonica, the dance was on. Folk tried to guess who was in the costumes. There was also the custom of bringing in the julebukk—a goat's head [usually made of wood or leather] on a stick.*[35]

Midsummer's Eve has been celebrated in Norway since ancient times. Long before it was declared a holy day in honor of John the Baptist (who was also known in Norway as St. Hans), it was a pagan rite that marked the time of the summer solstice or "turning of the sun." The sun shone longer on that day than any other time of the year; from that day until midwinter, the sun would shine less and less. It was an old European custom to light bonfires on Midsummer's Eve. Among other beliefs, this was thought to bring good luck and drive away evil spirits. The Scandinavians preserved the bonfire custom, often passing the night away by dancing around the leaping flames.[36]

In old Norway nearly every occasion in life was accompanied by dancing. If someone built a fishing boat, for instance, that was as good an excuse as any to invite some friends over, hire a fiddler, and dance far into the night.[37] People from Gudbrandsdal had what they called a *nystugudans* or "new-house dance" to celebrate the completion of a home.[38] Among the most important and memorable occasions were weddings.

From start to finish, Norwegian weddings offered a steady stream of music and dance, keeping everyone in high spirits. A fiddler led the bridal couple and their guests in a long procession to the church. Because there was not always room for everyone inside, some of the guests might dance out on the church lawn during the ceremony.[39] Following this, everyone followed the fiddler to the bride's home, where a feast awaited them. There the dancing would begin in earnest.

The bride needed to be light on her feet, since in most parts of Norway, she was expected to dance briefly with all the men present. Martin Bjørndal commented on this practice in an article about folk traditions in Møre:

> *After returning to the church, the bride had to dance with all the men. The oldest folks sat around the walls, watching and critiquing. They were reminded of their own dancing days.*[40]

In his book, *Excursions in the North of Europe in 1830 and 1833,* John Barrow gave an account of a country wedding in Hellesvit, Romsdal. He remarked that it was a local custom for the bride to do her first dance with one of the most distinguished men present—"usually someone of high esteem or a leading member of the church." At the particular wedding Barrow attended, it is interesting to note that this couple performed a waltz.[41]

A proper wedding celebration lasted at least three days. During that time, people danced both day and night, perhaps catching a few hours of sleep now and then.[42] The merry-making came to an end when the fiddler played a slow and melancholy tune called the *heimatslått* or "going-back-home tune." It was the signal for all the guests to thank their hosts and go their separate ways.[43]

Besides special occasions, there were many other opportunities to dance. Young people often met with a few friends on Sunday nights to talk and dance together.[44] A little more formal was the *kaffilag* or "coffee-group gathering." This was an evening of dancing and refreshments that was held at a particular farm. A certain group of neighbors from the surrounding community was invited to take part. In an article entitled, "Dans og Dansplassar," Andreas Mørch outlined some of the characteristics of a *kaffilag:*

> *Those who sponsored the gathering offered food and coffee to all the guests two times. They were responsible for obtaining a fiddler, but the men who came were supposed to pay him. The men also had to bring their own brandy. The kaffilag was held at a farm where there was room enough in the house to dance. It took a number of men and women to make up one of these gatherings. The dancing often went on all night.*[45]

The Norwegians did not always dance inside. During the summer months, they would display their skill on a level meadow or perhaps a bridge.[46]

Waltzing in the Nordfjord province.

Among Norway's folk dances, old-time couple dances are the most widely practiced today. A surprising number of Norwegians–probably the majority–have had an opportunity to try an old-time couple dance at one time or another. Some have learned the dances at workshops while others have learned them at home from family or friends. Old-time couple dances may be seen at folk-dance events as well as at some traditional events, such as Midsummer's Eve. They have become a familiar part of Norway's cultural heritage.

Notes

1 Anatole Chujoy, *The Dance Encyclopedia* (New York: A. S. Barnes and Company, 1949), s.v. "Old Time Dancing," p. 336.

2 Philip J. S. Richardson, *Social Dances of the Nineteenth Century in England* (London: Herbert Jenkins Ltd., 1960), p.44 Clarke and Crisp, *History of Dance,* p. 101.

3 Eduard Reeser, *The History of the Waltz* (Stockholm: The Continental Book Company, 1949), p. 1.

4 Chujoy, *Dance Encyclopedia,* s.v. "Waltz." Richardson, *Social Dances,* p. 44.

5 Patricia James, ed., *The Travel Diaries of Thomas Robert Matthus,* (Cambridge: Cambridge University Press, 1966), p. 26.

6 Salomo J. Wolf, *Beweis dass das Walzen eine Hauptquelle der Schwäche des Körpers und des Geistes unserer Generation sey* (Halle: 1799) quoted in Reeser, *History of the Waltz,* p. 21.

7 Gordon Yates, *The Ball: or a Glance at Almack's* (London: 1829).

8 A. P. Oppé and Cecil J. Sharp, *The Dance: An Historical Survey of Dancing in Europe* (London: Halton & Truscott Smith, Ltd., 1924), p. 30.

9 Thomas Wilson, *A Description of the Correct Method of Waltzing* (London: 1816).

10 Richardson, *Social Dances,* p. 42-44. Oppé and Sharp, *The Dance,* p. 29.

11 Clarke and Crisp, *History of Dance,* p. 102. Richardson, *Social Dances,* pp. 64, 77.

12 Oppé and Sharp, *The Dance,* p. 29.

13 Edward D. Clarke, *Travels in Scandinavia* (London: 1799), pp. 454-55.

14 Virgil L. Morton, *The Teaching of Popular Dance* (New York: J. Lowell Pratt & Company, 1966), p. 76.

15 Clarke and Crisp, *History of Dance,* p. 104.

16 Cellarius, *Fashionable Dancing* (London: 1847). M. Coulon, *The Ballroom Polka, Polka Cotillion, and Valse à Deux Temps* (London: 1844).

17 *Illustrated London News,* 11 May 1844.

18 Ibid.

19 Arthur Michel, "The Czech Polka," *Dance Magazine* (January-February 1944), quoted in Chujoy, *Dance Encyclopedia,* pp. 376-83.

20 Ballkort, *Handskriftsamlinga,* University of Oslo Library.

21 John Willis Clark and J. W. Dunning, *A Long Vacation Ramble in Norway and Sweden* (Cambridge, Mass.: Macmillan, 1857), p. 76.

22 Author's research in district of Sør-Trøndelag, December 1979; in district of Østerdal, July 1982.

23 G. Desrat, *Dictionnaire de la danse* (Paris: Librairies-imprimeries réunites, 1895).

24 Albert Zorn, *Grammatik der Tanzkunst* (Leipzig: 1887)–*Grammar of the Art of Dancing,* trans. Benjamin P. Coates, ed. Alfonso J. Sheafe (New York: Burt Franklin, 1905, reprint 1970), p. 233.

25 Harris, Pittman, and Waller, *Dance a While* (Minneapolis: Burgess Publishing Company, 1968), p. 233.

26 Richardson, *Social Dances,* p. 97.

27 Egil Bakka, *Norske Dansetradisjonar* (Oslo: Det Norske Samlaget, 1978), p. 109.

28 Mazurka step described in P. Valleau Cartier, *Waltz Instructor and Ballroom Guide* (New York: De Witt, Publisher, 1882), p. 48.

29 Zorn, *Grammatik,* s.v. "Polka-Mazurka."

30 Author's research in district of Valdres, July 1984.

31 Zorn, *Grammatik,* pp. 247-48.

32 Quoted in Richardson, *Social Dances,* p. 102.

33 Richardson, *Social Dances,* p. 118.

34 Clarke and Crisp, *History of Dance,* p. 105. The French name was *Pas de Quatre.* See Desrat, *Dictionnaire,* p. 270.

35 Anne Ødegård, *Liv og Leik in Valdres* (Gjøvik: Eigeforlag, Mariendals Boktrykkeri, 1945), p. 147.

36 Olav Bø, *Norske Års-Høgtider* (Oslo: Universitetsforlaget, 1980), pp. 108-115.

37 Brynjulf Alver and Arne Bjørndal,-*Og fela ho let* (Oslo: Universitetsforlaget, 1966), p. 66.

38 Olav Aukrust, "Tro og overtro omkring husflid og håndverk," *Håndverksliv* 1:302.

39 Hallvard Bergh, "Spel og dans i Valdres i eldre tid," *For Bygd og By* 16 (August 1916): 250.

40 Martin Bjørndal, *Segn og Tru: Folkeminne frå Møre* (Oslo: Norsk Folkeminnelag, Johansen & Nielsen, 1949), p. 113.

41 John Barrow, *Excursions in the North of Europe in 1830 and 1833* (London: John Murray, 1834), p. 310.

42 Peder Blom, *Vardal Prestegjeld* (Hamar: Norsk Skoletidendes Bogtrykkeri, 1899), p. 25. Alver and Bjørndal,-*Og fela,* p. 63.

43 Also known as *heimkomslått.* Alver and Bjørndal, — *Og fela,* p. 63.

44 Thora Skolmen, "Leik og Dans," *Norsk Folkekultur: Folkeminne Tidskrift* (Skien: Bødr. Dyrings Prenteverk, 1935), p. 88.

45 Andreas Mørch, "Dans og Dansarplassar," *Folkeminne frå Sigdal og Eggedal* (Oslo: Johansen & Nielsen, 1932), pp. 127-28.

46 Ibid., pp. 120-21.

Byte reinlender

Reinlender Mixer

Formation: Several couples in circle B. Partners face CCW LOD with following position:

Steps: Reinlender step, step-hop

MUSIC 4/4

Phrase	Ms	Part
		I. FORWARD; COUPLE TURNS CW
A	1-2	With man beginning left foot and woman right, perform 2 reinlender steps in CCW LOD. On last ct, man turns a little toward partner.
	3-4	Assume position 2. With man beginning left foot and woman right, make 2 CW turns with 4 step-hops (2 step-hops per turn). On last ct, partners open away from each other to face CCW LOD with side-by-side position, as shown above.
	5-8	Repeat action of ms 1-4.
		II. MAN MOVES AHEAD TO NEXT WOMAN; COUPLE TURNS CW Release position with partner.
	9-10	Beginning left foot, man performs 2 reinlender steps forward to next woman in circle. As he comes beside her, they assume side-by-side position. On last ct, man turns a little toward his new partner.
		Beginning right foot, woman performs 2 reinlender steps, traveling only a little forward in CCW LOD. She assumes side-by-side position with new partner.
		Note: In order to perform this part, couples must be evenly spaced around circle and not more than about 6 feet apart from each other.
	11-12	Assume position 2. With man beginning left foot and woman right, make 2 CW turns with 4 step-hops (2 step-hops per turn). On last ct, partners open away from each other to face CCW LOD with side-by-side position.
	13-16	Repeat action of ms 1-4, part I.

* * *

Repeat part II for remainder of dance. Each man will move ahead to next woman in circle, turn with her, and then dance and turn with his new partner once more.

Familie vals

Family Waltz

Formation: 4 or more couples in circle A, facing center of circle with light handhold.

Steps: Step-swing, waltz

MUSIC 3/4 ______

Phrase	Ms	Part	
		I.	**STEP-SWINGS IN PLACE**
A	1		Stepping onto left foot, man performs 1 step-swing in place, turning a little toward corner person and acknowledging her.
			Stepping onto right foot, woman performs 1 step-swing in place, turning a little toward corner person and acknowledging him.
	2		Stepping onto right foot, man performs 1 step swing in place, turning a little toward partner and acknowledging her.
			Stepping onto left foot, woman performs 1 step-swing in place, turning a little toward partner and acknowledging him.
	3-4		Repeat action of ms 1-2.
			Note: During above figure, hands remain lifted with light handhold.
		II.	**WOMAN ACROSS; WALTZ**
			Release handhold with partner (man's right hand, woman's left).
	5		Beginning left foot, man performs 1 waltz step in place. At same time, he leads corner person in front of him with left hand and assumes position 1 with her.
			Beginning right foot, woman performs 1 waltz step, coming in front of corner person and assuming position 1 with him.
	6-8		With man beginning right foot and woman left, make 1 CW turn with 3 waltz steps. With last waltz step, man and woman open away from each other, such that woman concludes on right side of man in circle.
			Join hands in circle with light handhold. You should now have a new partner and corner person. Note that this dance is a mixer where women progress to a new partner each time above figures are done.

REPEAT DANCE FROM BEGINNING

Gammel reinlender

Old Reinlender

Formation: Several couples in circle B. Partners face CCW LOD and join inside hands with light handhold. Free hand is on hip, thumb forward.

Steps: Reinlender step, step-hop

Note: This dance has a 2/4 rhythm. To avoid confusion with numerous eighth notes, the dance has been described below as if it had a 4/4 rhythm, making it possible to count 1-2-3-4 per measure instead of 1-&-2-&. The total number of measures will remain the same.

MUSIC 4/4

Phrase Ms Part

I. FORWARD

A 1-2 With man beginning left foot and woman right, perform 1 reinlender step in CCW LOD. Following this, man and woman turn toward each other and dance 1 reinlender step in CW LOD, such that their footwork travels sideways:

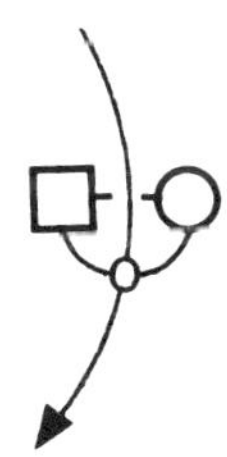

3-4 With man beginning left foot and woman right, turn to face CCW LOD and dance 4 step-hops forward.

5-8 Repeat action of ms 1-4.

II. FORWARD; COUPLE TURNS CW

Partners assume position 1. Man and woman are angled toward CCW LOD:

9-10 With man beginning left foot and woman right, perform 1 reinlender step in CCW LOD. Without releasing position, reverse footwork and perform 1 reinlender step in CW LOD.

11-12 With man beginning left foot and woman right, make 2 CW turns with 4 step-hops (2 step-hops per turn).

13-16 Repeat action of ms 9-12, part II.

III. WOMAN TURNS CCW

Partners face CCW LOD and join inside hands with light handhold. Free hand is on hip, thumb forward.

B	1-2	Repeat action of ms 1-2, part I.
	3-4	Beginning left foot, man performs 4 step-hops forward. At same time, he lifts right hand such that woman can turn beneath it. Beginning right foot, woman makes 2 CCW turns beneath uplifted arms with 4 step-hops. Footwork travels in CCW LOD.
	5-8	Repeat action of ms 1-4, part III.

IV. WOMAN TURNS CW

Partners face CCW LOD and join inside hands with light handhold. Free hand is on hip, thumb forward.

	9-10	Repeat action of ms 1-2, part I.
	11-12	Beginning left foot, man performs 4 step-hops forward. At same time, he swings woman's left hand forward and lets go, so that she can turn away. He then crosses arms above chest and parallel to ground. Beginning right foot, woman makes 2 CW turns by herself with 4 step-hops. Footwork travels in CCW LOD. Both hands are placed on hips, thumbs forward.
	13-16	Repeat action of ms 9-12, part IV, but with one change. During woman's turn, she will make only 1½ CW turns, such that woman concludes facing CW LOD. Woman performs 3 step-hops plus two small steps (left-right) so that she can begin next figure on left foot.

V. HALF-TURNS IN PLACE

Partners assume position 5.

A	1-2	Beginning left foot, dance 3 small step-hops forward, such that partners make ½ CW turn in place. On next ct, swing right foot forward such that heel touches floor with toes pointing up. Upper body tilts a little backward as leg swings forward. Right leg is straight; supporting left leg is bent somewhat.
	3-4	Beginning right foot, dance 3 small step-hops backward, such that partners make ½ CCW turn in place. On next ct, swing left foot forward such that heel touches floor with toes pointing up. Upper body tilts a little backward as leg swings forward. Left leg is straight; supporting right leg is bent somewhat.
	5-12	Repeat action of ms 1-4, part V, two times.
	13-14	Repeat action of ms 1-2, part V.

Part III.

Part V.

Part VI.

Part VII.

15 Beginning right foot, each person performs 2 step-hops backward. Release handhold.

16 On cts 1-2-3 of next ms, man stamps 3 times in place with right-left-right. He should conclude facing CCW LOD with weight on both feet.

At same time, woman releases handhold and makes ½ CW turn with 3 steps in place, right-left-right. She should conclude facing CCW LOD with weight on right foot.

Note: Woman's turn is described according to directions in Klara Semb's book, *Norske Folkedansar.* A common variation, however, is for woman to release only right hand and make ½ CCW turn in place with 3 steps in place, right-left-right.

VI. CROSS TO SIDE OF PARTNER

Man crosses arms above chest and parallel to ground. Woman places both hands on hips, thumbs forward.

B 1 — Beginning right foot, man performs 1 reinlender step, moving sideways to right and passing behind woman.

Beginning left foot, woman performs 1 reinlender step, moving sideways to left and passing in front of man.

2 — Reverse roles of ms 1, such that man passes in front of woman and woman behind man.

3-4 — Join hands (man's right hand, woman's left) with light handhold and place free hand on hip, thumb forward. With man beginning right foot and woman left, perform 4 step-hops forward.

15-16 — Repeat action of ms 1-4, part VI, three times, but with one change. Instead of performing 4 step-hops during ms 3-4 of third repeat, woman performs 3 step-hops plus 2 small steps (right-left) so that she can begin next figure on right foot.

VII. MAN KNEELS; WOMAN AROUND HIM

A 1-2 — Without releasing handhold, man quickly drops onto right knee, with left knee up. His back should be straight. Man raises right arm directly upward and holds woman's left hand.

Beginning right foot, woman makes 1 CCW circle around man with 2 reinlender steps. Partners look at each other as woman dances around man.

3-4 — On ct 1, man quickly rises to dance 4 step-hops forward, beginning left foot. He continues to hold woman's left hand with his right, now with light handhold .

Beginning right foot, woman performs 4 step-hops forward.

5-16 — Repeat action of ms 1-4, part VII, three times.

VIII. SPLIT AWAY; COME TOGETHER; TURN

Man crosses arms above chest and parallel to ground. Woman places both hands on hips, thumbs forward.

B 1 — Beginning left foot, man performs 1 reinlender step toward center of circle.

Beginning right foot, woman performs 1 reinlender step away from center of circle.

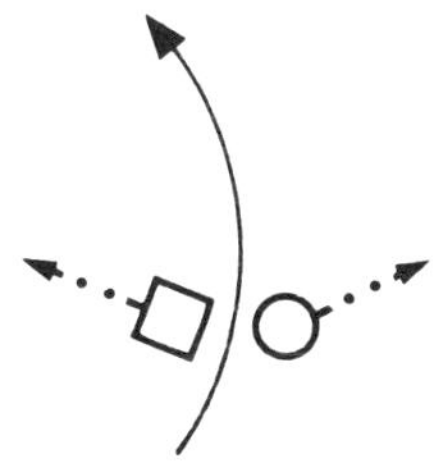

2 Beginning right foot, man turns a little CW and performs 1 reinlender step toward partner. He stamps on first step. While dancing toward partner, man unfolds arms, extends them outward, and claps on ct 3.

Beginning left foot, woman turns a little CCW and performs 1 reinlender step toward partner.

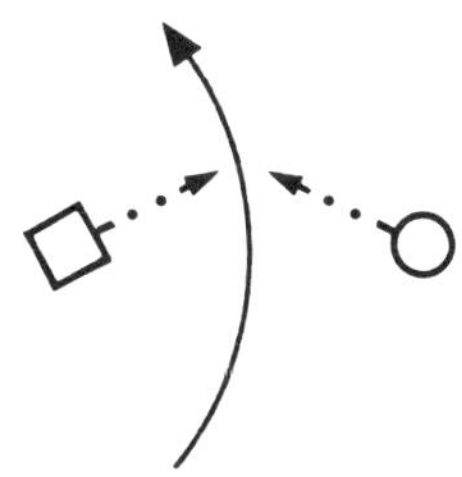

3-4 Partners assume position 1 and make 2 CW turns with 4 step-hops.

5-16 Repeat action of ms 1-4, part VIII, three times.

IX. COUPLE TURNS CW; LIFT

Partners assume following position:

Woman's right arm is down by her side and rotated outward, such that inner side of arm faces partner. Right hand makes a fist. Her left hand is centered firmly on man's right shoulder.

Man's left arm is down by his side and rotated outward, such that inner side of arm faces partner. He supports woman's right fist from below with his left hand. He places right hand around woman's shoulder blade area.

A 1-2 With man beginning left foot and woman right, perform 1 reinlender step sideways in CCW LOD. Without releasing position, reverse footwork and perform 1 reinlender step in CW LOD.

3 With man beginning left foot and woman right, make about 1½ CW turns with 2 step-hops. Extra half turn means that couple must rotate quicker than usual. Man should conclude facing toward center of circle, woman away.

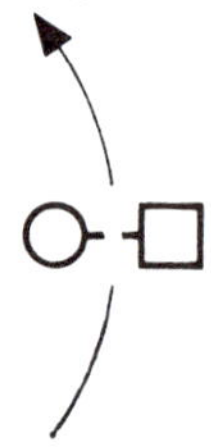

4 On cts 1-2, man brings feet together and lifts woman (with good flexion in knees), making about ½ CW turn with her in air. On cts 3-4, he puts woman down on other side.

Woman prepares for lift by bending knees a little and then by jumping vertically on ct 1. As woman jumps, her right arm should be held stiffly so that man can lift her. She also presses down on man's shoulder for extra lift and support. While woman is in air, legs should be together. Woman lands with flexion in ankles and knees.

Center of circle 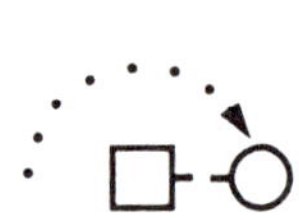

5-8 Repeat action of ms 1-4, part IX.

Note: Gammel reinlender can be made to fit most reinlender recordings by doing one or more figures 2 times instead of 4 times.

Hoppvals

Hopping-waltz
Røros, Sør-Trøndelag

Formation: Several couple in circle A. Partners assume following position:

Man places right hand around woman's shoulder-blade area. He holds fingers of woman's right hand with his left and presses that hand against left side of his waist, such that his thumb is forward. His left arm is comfortably bent.

Woman places left hand on man's right shoulder. She lets man take her right hand and hold it by his waist.

Man faces CCW LOD, woman CW LOD.

Steps: Hoppvals step

MUSIC 2/4 ________________________________

Phrase	Ms	Part	
		I.	**TRAVEL IN CCW LOD**
A	1-3		With man beginning left foot and woman right, perform 3 hoppvals steps in CCW LOD. Man dances forward, woman backward.
		II.	**COUPLE TURNS CW**
	4-16		With man beginning right foot and woman left, make 6 CCW turns with 12 hoppvals steps (2 hoppvals steps per turn). Following this, perform 1 more hoppvals step mostly in place. This last step helps couple to stop momentum of their turn. Man should conclude facing CCW LOD, woman CW LOD.
		III.	**TRAVEL IN CCW LOD**
B	1-4		With man beginning left foot and woman right, perform 4 hoppvals steps in CCW LOD. Man dances forward, woman backward.
		IV.	**COUPLE TURNS CW**
	5-16		With man beginning left foot and woman right, make 6 CW turns with 12 hoppvals steps (2 hoppvals steps per turn). Man should conclude facing CCW LOD, woman CW LOD.

REPEAT DANCE FROM BEGINNING

Lettisk polka

Latvian Polka*

Formation: Several couples in circle B. Partners face CCW LOD with following position:

Steps: Change step, heel-toe (explained below), light step

*This appears to be a Norwegian dance with some Latvian characteristics.

MUSIC 2/4

Phrase	Ms	Part	
		I.	**HEEL-TOE** Place weight on right foot.
A	1		On ct 1, perform a small hop on right foot. At same time, quickly touch your left heel on floor to left diagonal. On ct 2, perform another small hop on right foot. At same time, quickly touch toe of left foot beside ball of right foot.
	2		On ct 1, perform a small hop on right foot. At same time, quickly touch your left heel on floor to left diagonal. On ct 2, perform a small hop on right foot and make about ½ CW turn in place, such that man and woman turn individually. At same time, bring left foot beside right, landing with weight on both feet.
			Do not release handhold during turn. Man extends left arm behind woman's shoulders and bends right arm. Woman extends right arm in front of man's chest and bends left arm. This position should be opposite of previous position.
	3		Place weight on left foot and perform heel-toe action with right foot.
	4		On ct 1, perform a small hop on left foot. At same time, quickly touch your right heel to right diagonal. On ct 2, perform a small hop on left foot and make about ½ CCW turn in place, such that man and woman turn individually. At same time, bring right foot beside left, landing with weight on both feet.
			Do not release handhold during turn. Assume opposite arm positions.
		II.	**FORWARD; WOMAN TURNS**
	5-6		Beginning left foot, perform 2 change steps forward in CCW LOD.
	7-8		Release left hands.
			Beginning left foot, man performs another 2 change steps forward. At same time, he lifts his right hand, such that woman can turn beneath it. His left arm is relaxed by his side.

Beginning left foot, woman makes 2 CW turns beneath uplifted arms with 4 light steps. Footwork travels in CCW LOD. She holds skirt to side with left hand.

Note: Woman turns quickly, pivoting with each light step. There are 2 light steps per turn. Woman performs first of these by rotating her left foot inward, such that she turns her back to line of direction. She performs second light step by rotating her right foot outward, such that she turns to face line of direction. It is possible to omit this turning now and then to avoid getting dizzy. To do this, simply perform 2 change steps forward as before.

REPEAT DANCE FROM BEGINNING

To begin dance again, quickly assume beginning position. Woman must stop her turning momentum and be ready to touch left heel on floor.

Parisarpolka

Parisian Polka

Western Norway

Formation: Several couples in circle B. Man and woman are angled toward CCW LOD. Inside hands are joined with light handhold. Free hand is on hip, fingers forward.

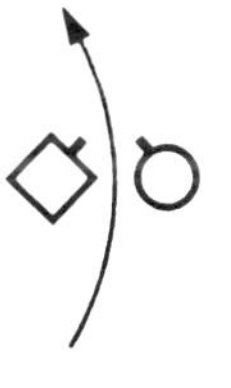

Steps: Change step, light step, pivot step*

* To perform a pivot step, take a step forward, bending somewhat at ankle and knee, and then rise such that you turn quickly on ball of foot.

MUSIC 2/4

Phrase	Ms	Part
		I. FORWARD AND REVERSE; REFRAIN
A	1-2	With man beginning left foot and woman right, perform 3 light steps in CCW LOD on cts 1-2-1. Following this, extend free leg forward on ct 2 and touch toe on floor (see illus.).

3-4 Without releasing handhold, turn and repeat action of ms 1-2 in opposite direction with opposite footwork.

5-6 Partners turn to face CCW LOD without releasing handhold. With man beginning left foot and woman right, perform 2 change steps in CCW LOD.

7-8 *Refrain.* Partners assume position 3. With man beginning left foot and woman right, make 2 CW turns with 4 pivot steps (2 pivot steps per turn). Footwork is interlocking, such that man and woman step between partner's feet with right foot. As couple turns CW, they should also progress a little forward in overall CCW LOD.

9-16 Repeat action of ms 1-8.

II. TURN AWAY; REFRAIN

Release handhold, relaxing arms by your sides.

B 1-2 With man beginning left foot and woman right, partners make 1 turn away from each other (man CCW, woman CW) with 3 light steps on cts 1-2-1. Footwork travels in CCW LOD. Following this, extend free leg forward on ct 2 and touch toe on floor.

3-4 Repeat action of ms 1-2, part II in opposite direction with opposite footwork.

5-6 Arms remain relaxed to either side. Partners turn to face CCW LOD. With man beginning left foot and woman right, perform 2 change steps in CCW LOD.

7-8 *Refrain.* Repeat action of ms 7-8, part I.

9-16 Repeat action of ms 1-8, part II.

III. TURN AWAY UNDER ARMS; REFRAIN

Partners join hands across with low handhold.

A 1-2 With man beginning left foot and woman right, partners make 1 turn away from each other (man CCW, woman CW) with 3 light steps on cts 1-2-1.

Footwork travels a little forward in CCW LOD. At start of turn, swing hands toward CCW LOD and raise above head level as you come back-to-back with partner. Conclude facing partner with low handhold. On next ct, touch toe of free foot on floor in place.

3-4 Repeat action of ms 1-2, part III in opposite direction with opposite footwork. Note that hands will now swing toward CW LOD and raise above head level.

5-6 Without releasing handhold, lift hands to about waist level. Partners turn to face CCW LOD. With man beginning left foot and woman right, perform 2 change steps in CCW LOD.

7-8 *Refrain.* Repeat action of ms 7-8, part I.

9-16 Repeat action of ms 1-8, part III.

IV. SIDE POSITION; WOMAN TURNS CW; REFRAIN

Partners assume following position:

Woman immediately turns on supporting left foot to face CW LOD. She extends right arm to side and holds man's left hand with her right. She bends left arm, lifting elbow a little, and holds man's right hand with her left.

Man faces CCW LOD. He is to left side of woman and assumes same arm and hand positions as given above.

B 1-2 With man beginning left foot and woman right, perform 3 light steps in CCW LOD on cts 1-2-1. Woman travels backward. Following this, touch toe of free foot on floor in place.

3-4 Without releasing handhold, woman makes ½ CCW turn and man makes ½ CW turn to face opposite direction. This is performed as follows:

Man lifts right hand above woman's head and draws it to his right side, bending right arm and lifting elbow a little as he does so. At same time, he lowers and extends his left arm behind woman's back.

Woman passes left hand over head and extends left arm to side. She bends right arm and places back of right hand against her lower back.

With man beginning left foot and woman right, perform 3 light steps in CW LOD on cts 1-2-1; above turn happens mostly on first step. Woman travels backward with remaining steps. Following this, touch toe of free foot on floor in place.

5-6 Man releases woman's right hand with his left, but continues to hold her left hand with his right. He makes ½ CCW turn to face CCW LOD. Beginning left foot, man dances 2 change steps forward. At same time, he lifts right arm such that woman can turn beneath it. His left arm is relaxed by his side.

Woman makes 2 CW turns beneath man's uplifted arm with 4 pivot steps, right-left-right-left. Pivot steps travel in CCW LOD. Woman holds skirt to side with right hand.

7-8 As woman completes last turn, partners quickly assume position 3 and dance refrain. Repeat action of ms 7-8, part I.

9-16 Repeat action of ms 1-8, part IV. Note that position no. 3 opens immediately into above side position to repeat figure.

V. WOMAN TURNS UNDER ARM; REFRAIN

Man and woman are angled toward CW LOD. Inside hands are joined with light handhold.

A 1-2 Beginning left foot, man performs 3 light steps in CCW LOD on cts 1-2-1. At same time, he lifts right hand such that woman can turn beneath it. Following this he extends free leg forward on ct 2 and touches toe on floor.

Beginning right foot, woman makes 1 CW turn beneath man's uplifted arm with 3 light steps. Footwork travels in CCW LOD. Following this, she extends free leg forward on ct 2 and touches toe on floor.

3-4 Without releasing handhold, turn and repeat action of ms 1-2, part V, in opposite direction with opposite footwork. Woman now makes 1 CCW turn.

5-6 Beginning left foot, man dances 2 change steps forward. At same time, he lifts right arm such that woman can turn beneath it. His left arm is relaxed by his side.

Woman makes 2 CW turns beneath man's uplifted arm with 4 pivot steps, right-left-right-left. Pivot steps travel in CCW LOD. Woman holds skirt to side with right hand.

7-8 As woman completes last turn, partners quickly assume position 3 and dance refrain. Repeat action of ms 7-8, part I.

9-18 Repeat action of ms 1-8, part V.

Reinlender

Formation: Several couples in circle B. Partners face CCW LOD with following position:

Steps: Reinlender step, step-hop

MUSIC 2/4

Phrase	Ms	Part
		I. FORWARD; COUPLE TURNS CW
A	1-2	With man beginning left foot and woman right, perform 2 reinlender steps in CCW LOD. On last ct, man turns a little toward partner.
	3-4	Assume position 2. With man beginning left foot and woman right, make 2 CW turns with 4 step-hops (2 step-hops per turn). On last ct, partners open away from each other to face CCW LOD with side-by-side position, as shown above.

REPEAT DANCE FROM BEGINNING

* * *

VARIATION

After performing above, partners may substitute action of ms 1-2 with several other figures. One of these is given below.

Phrase	Ms	
A	1	With man beginning left foot and woman right, partners make 1 turn away from each other (man CCW, woman CW) with 1 reinlender step. Footwork travels in CCW LOD. On last ct, man may slap side of his right shoe with right hand. He does this by quickly lifting right foot toward hand.
	2	With man beginning right foot and woman left, perform 1 reinlender step forward. At same time, partners come together and assume side-by-side position. On last ct, man turns a little toward partner.
	3-4	Repeat action of ms 3-4.

Røros ringlender

Røros, Sør-Trondelag

Formation: Several couples in circle B. Partners face CCW LOD with light handhold. Free hand is on hip, thumb forward.

Steps: Reinlender step, ringlender step, step-hop, trio step

Note: This dance has a 2/4 rhythm. To avoid confusion with numerous eighth notes, the dance has been described below as if it had a 4/4 rhythm, making it possible to count 1-2-3-4 per measure instead of 1 & 2 &. The total number of measures will remain the same.

MUSIC Phrase	4/4 Ms	Part	
		I.	**FORWARD WITH RINGLENDER**
A	1-3		With man beginning left foot and woman right, perform 2 ringlender patterns in CCW LOD.
	4		With man beginning left foot and woman right, perform first 3 steps of another ringlender pattern (step, close, step-hop) in CCW LOD.
		II.	**PARTNERS COME TOGETHER AND TURN CW**
	5-8+		Man comes toward woman with 1 step-hop on right foot. He stamps as he takes his step.
			At same time, woman performs 1 step-hop on left foot.
			Partners assume position 3. With man beginning left foot and woman right, make 3½ CW turns with 7 step-hops* (2 step-hops per turn). Still turning, perform 1 additional step, such that man concludes facing CCW LOD, woman CW LOD. This last step takes place on *first 2 cts of next musical phrase.*

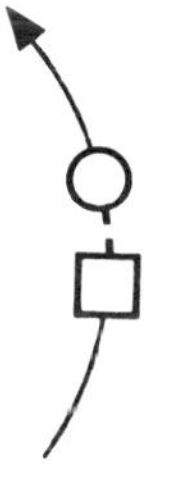

*Note: Some traditional dancers in Røros perform a smooth trio step instead of a step-hop. Others combine these steps, performing 1 trio step and then 1 step-hop or vice versa.

MUSIC Phrase	4/4 Ms	Part	
		III.	**COUPLE TURNS CCW**
B	1-8		Partners are still in position 3. With man beginning left foot and woman right, perform 1 reinlender step in CCW LOD, such that man dances forward, woman backward. Turn a little CCW with last step in preparation for next action.

Note: Since first 2 cts of phrase B were borrowed for last step of part II, this part begins on ct 3. Perform reinlender step on cts 3-4-1-2.

With man beginning right foot and woman left, make 3 CCW turns with 6 reinlender steps (2 reinlender steps per turn). You must turn a little with each step of reinlender pattern in order to come completely around. Footwork travels in CCW LOD. Partners may rock a little from side to side as they perform reinlender steps.

Note: Perform each reinlender step on cts 3-4-1-2.

On cts 3-4 of last ms, partners open away from each other with 1 step (man right foot, woman left). They face CCW LOD and join inside hands (man's right hand, woman's left) with light handhold.

REPEAT DANCE FROM BEGINNING

Seattle Polka

Formation: Several couples in circle B. Partners face CCW LOD and join inside hands with light handhold. Free arm is relaxed by side.

Steps: Pivot step*, walking step

* To perform a pivot step, take a step forward, bending somewhat at ankle and knee, and then rise such that you turn quickly on ball of foot.

This is a Scandinavian-American dance, probably descended from a Norwegian parisarpolka.

MUSIC 2/4

Phrase	Ms	Part
		I. FORWARD AND REVERSE
A	1-2	With man beginning left foot and woman right, perform 3 walking steps in CCW LOD on cts 1-2-1. On next ct, pivot ½ turn (man CW, woman CCW) on supporting foot to face opposite direction. A common variation is to perform a slight hop as you turn and then stamp foot on floor (on ct 2) as you come down.
	3-4	Partners assume light handhold with other hand (man's left, woman's right). With man beginning right foot and woman left, partners perform 3 walking steps in CW LOD on cts 1-2-1. On next ct, pivot ½ turn (man CCW, woman CW) on supporting foot to face opposite direction. Clap hands together 2 times on cts &-2 as you turn.

II. COUPLE TURNS CW

5-6 Partners assume following position:

With man beginning left foot and woman right, perform 4 walking steps in CCW LOD. On last step, man turns a little toward partner.

7-8 Partners assume position 2. With man beginning left foot and woman right, make 2 CW turns with 4 pivot steps (2 pivot steps per turn). Footwork is interlocking, such that man and woman step between partner's feet with right foot. As couple turns, they should also progress a little forward in CCW LOD.

REPEAT DANCE FROM BEGINNING

VARIATION

Whenever they wish to do so, partners may substitute action of ms 1-4 with following variation.

I. TURN AWAY

A 1-2 Release handhold. With man beginning left foot and woman right, partners make ¾ turn away from each other (man CCW, woman CW) with 3 steps on cts 1-2-1. Footwork travels in CCW LOD. Conclude facing partner. On next ct, touch ball of free foot beside ball of supporting foot.

3-4 With man beginning right foot and woman left, partners make 1 turn away from each other (man CW, woman CCW) with 3 steps on cts 1-2-1. Footwork travels in CW LOD. Conclude facing partner. On next ct, touch ball of free foot beside ball of supporting foot.

Note: Partners may perform their turns in a rather free manner. One partner, for instance, may turn while the other dances forward. It is possible to turn on ms 1-2, but not on ms 3-4, and vice versa. A stamp may also be used at the end of the first turn and 2 claps at the end of the second.

Springpolka

Springing-polka*

Formation: Several couples in circle B. Partners face CCW LOD and join inside hands with light handhold. Man relaxes left arm by his side. Woman holds skirt to side with right hand.

Steps: Change step (3/4 rhythm, first version), step-swing, springpolka

*As performed in Nordfjord region

MUSIC 3/4 ______________________________

Phrase	Ms	Part	
		I.	**STEP-SWINGS IN PLACE**
A	1		Stepping onto left foot, man performs 1 step-swing in place.
			Stepping onto right foot, woman performs 1 step-swing in place.
	2		Repeat action of ms 1 with opposite footwork.
		II.	**FORWARD**
	3-4		With man beginning left foot and woman right, perform 2 change steps forward. On last ct, man turns a little toward partner.
		III.	**SPRINGPOLKA**
			Partners assume position 2.
	5-8		With man beginning left foot and woman right, make 4 CW turns with 4 springpolka steps. On last ct, partners open away from each other to face CCW LOD.

REPEAT DANCE FROM BEGINNING.

Sunnmørsreinlender

Sunnmøre Reinlender

Formation: Several couples in circle B. Partners face CCW LOD and join hands with light handhold. Free hand is on hip, thumb forward.

Steps: Change step, light running step, reinlender step, step-hop, trio step

Note: This dance has a 2/4 rhythm. To avoid confusion with numerous eighth notes, the dance has been described below as if it had a 4/4 rhythm, making it possible to count 1-2-3-4 per measure instead of 1-&-2-&. The total number of measures will remain the same.

MUSIC 4/4

Phrase	Ms	Part
		I. FORWARD
A	1-2	With man beginning left foot and woman right, perform 2 reinlender steps in CCW LOD.
	3-4	With man beginning left foot and woman right, continue traveling forward with 4 step-hops.
	5-8	Repeat action of ms 1-4.
		II. FORWARD; COUPLE TURNS CW Partners assume position 1. Man and woman are angled toward CCW LOD:

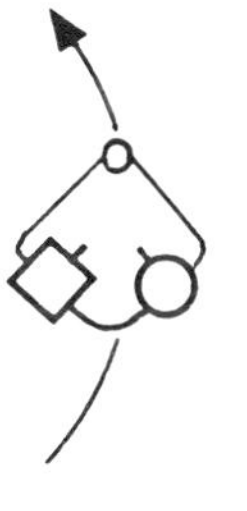

Phrase	Ms	Part
	9-10	With man beginning left foot and woman right, perform 2 reinlender steps in CCW LOD.
	11-12	With man beginning left foot and woman right, make 2 CW turns with 4 trio steps (2 trio steps per turn).
	13-16	Repeat action of ms 9-12, part II.
		III. WOMAN TURNS CW; MAN TURNS CCW; COUPLE TURNS CW Man continues to hold woman's right hand with his left. Man lowers right arm by his side, and woman lowers left arm by her side.
B	1	Beginning left foot, man performs 1 reinlender step in CCW LOD. At same time, he lifts left hand such that woman can turn beneath it.

Beginning right foot, woman makes 1 CW turn beneath uplifted arms with 1 reinlender step. Footwork travels in CCW LOD.

2 Beginning right foot, man makes 1 CCW turn beneath uplifted arms with 1 reinlender step. Footwork travels in CCW LOD.

Beginning left foot, woman performs 1 reinlender step in CCW LOD. At same time, she lifts right hand such that man can turn beneath it.

3-4 Partners assume position 1. With man beginning left foot and woman right, make 2 CW turns with 4 trio steps (2 trio steps per turn).

5-16 Repeat action of ms 1-4, part III, three times.

IV. TURN AWAY; FORWARD; COUPLE TURNS CW

Man crosses arms above chest and parallel to ground. Woman places both hands on hips, thumbs forward.

C 1 With man beginning left foot and woman right, partners make 1 turn away from each other (man CCW, woman CW) with 1 reinlender step. Footwork travels in CCW LOD.

2 With man beginning right foot and woman left, perform 1 reinlender step forward. At same time, partners come beside each other and join inside hands with light handhold. Free hand is relaxed by side.

3-4 Partners assume position 1. With man beginning left foot and woman right, make 2 CW turns with 4 trio steps (2 trio steps per turn).

5-16 Repeat action of ms 1-4, part IV, three times.

V. ARM-CROSS; CHANGE PLACES

Partners face CCW LOD. Join both hands (left hands together, right hands together) such that man's right arm crosses behind woman's left arm.

A 1 Beginning left foot, man performs 3 steps, moving a little sideways to his right (first step crosses behind right foot). At same time, he leads woman in front of him with hands. Man then touches right heel on floor to right diagonal.

Beginning right foot, woman passes in front of man and concludes on his left side with 3 light running steps. Do not release handhold. By end of ms, woman should be facing CCW LOD. Woman touches left heel on floor to left diagonal.

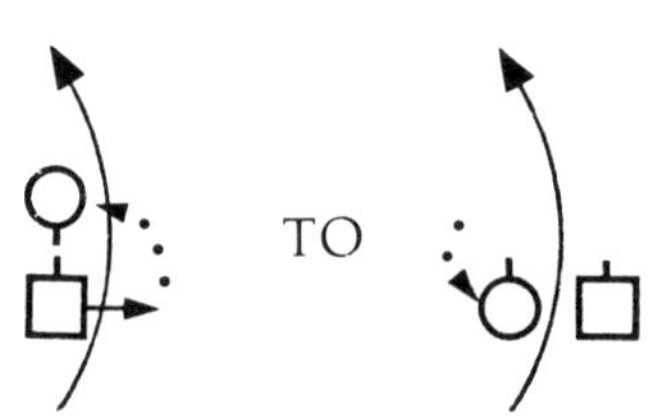

2 Repeat action of ms 1, part V, in opposite direction with opposite footwork.

3-4 Partners assume position 1. With man beginning left foot and woman right, make 2 CW turns with 4 trio steps (2 trio steps per turn).

5-16 Repeat action of ms 1-4, part V, three times.

VI. MAN KNEELS; WOMAN AROUND HIM

B 1-2 Man quickly drops onto right knee, with left knee up. His back should be straight. Man raises right arm directly upward and holds woman's left hand with his right. Free hand is on hip, thumb forward.

Beginning right foot, woman makes 1 CCW circle around man with 2 reinlender steps. Partners look at each other as woman dances around man. Her free hand is on hip, thumb forward.

3-4 On ct 1, man quickly rises to dance 4 step-hops forward, beginning left foot. He continues to hold woman's left hand with his right, now with light handhold.

Beginning right foot, woman performs 4 step-hops forward.

5-16 Repeat action of ms 1-4, part VI, three times.

VII. SPLIT AWAY; COME TOGETHER; LIFT

Man crosses arms above chest and parallel to ground. Woman places both hands on hips, thumbs forward.

C 1 Beginning left foot, man performs 1 reinlender step toward center of circle.

Beginning right foot, woman performs 1 reinlender step away from center of circle.

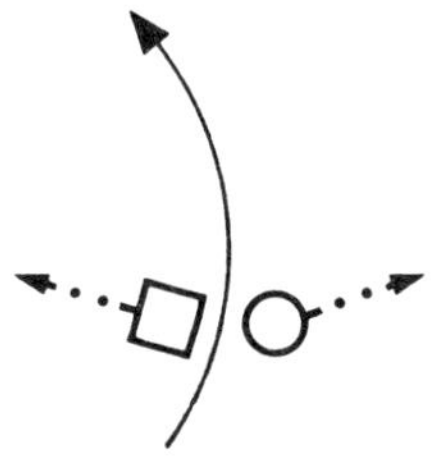

2 Beginning right foot, man turns a little CW and performs 1 reinlender step toward partner. He stamps on first step. While dancing toward partner, man unfolds arms, extends them outward, and claps on ct 3.

Beginning left foot, woman turns a little CCW and performs 1 reinlender step toward partner.

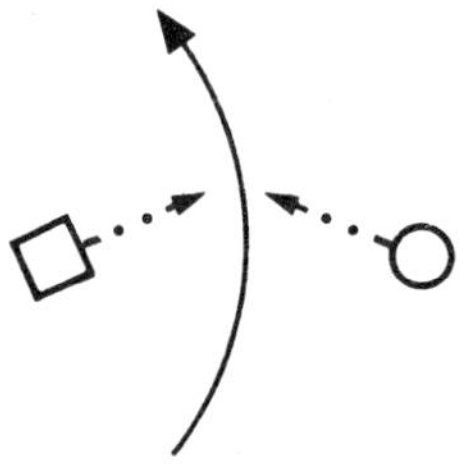

5-16 Repeat action of ms 1-4, part VII, three times.

Note: Sunnmørsreinlender can be made to fit most reinlender recordings by doing one or more figures 2 times instead of 4 times.

Music

Sju gjenter går i ringen

Seven Girls Walk in the Ring

Traditional

Sju gjenter går i ringen,*
så glade som den lette vind.
For det er nettopp tingen,
å finne venen sin.
Hei, hopp-sa-sa, falleral-la-la!
Hei, hopp-sa-sa, falleral-la-la!
For det er nettopp tingen
å finne venen sin.
Hei, hopp-sa-sa, falleral-la-la!
Hei, hopp-sa-sa, falleral-la-la!
For det er nettopp tingen
å finne venen sin.

Seven girls walk in the ring,
cheerful as a summer breeze.
It is really fun
to find some friends.
Hei, hopp-sa-sa, falleral-la-la!
Hei, hopp-sa-sa, falleral-la-la!
It is really fun
to find some friends.
Hei, hopp-sa-sa, falleral-la-la!
Hei, hopp-sa-sa, falleral-la-la!
It is really fun
to find some friends.

*May also be sung, "Sju guter går i ringen" (Seven boys walk in the ring).

Å jul med din glede

O Christmas with Joy

Gustava Kjelland

1. Å jul med din glede og barnlige lyst,
vi ønsker deg alle velkommen.
Vi hilser deg alle med jublende røst,
titusende ganger velkommen!

Refrain:
Vi klapper i hendene,
vi synger og vi ler,
så glad er vi, så glad er vi.
Vi svinger oss i kretsen og neier
og bukker.

2. I østerlands vise, i tre stjernemenn,
vi vett jo nok hvorhen i drage.
Ti vi ville også så gjerne derhen,
og eder på reisen ledsage.

Refrain

3. Så rekker jeg deg nå med glede min hånd,
kom skynd deg og gi meg den andre.
Så knytter vi kjærleikens hellige bånd
og lover å elske hverandre.

Refrain

1. O Christmas with joy and childlike delight,
we wish you our heartiest welcome.
We greet you with cheerful songs,
and bid you welcome a thousand times!

Refrain:
We clap our hands,
we sing and laugh,
so glad are we, so glad are we.
We spin around in a circle and curtsey
and bow.

2. From the land of the East, came three wisemen;
we know what place they were seeking.
We wish we could be there
and accompany them on their journey.

Refrain

3. With gladness I reach out my hand,
now hurry, and give me yours.
Then we will tie the bonds of friendship,
and promise to love each other.

Refrain

Jeg gikk meg over sjø og land

I Traveled Over Sea and Land

Traditional

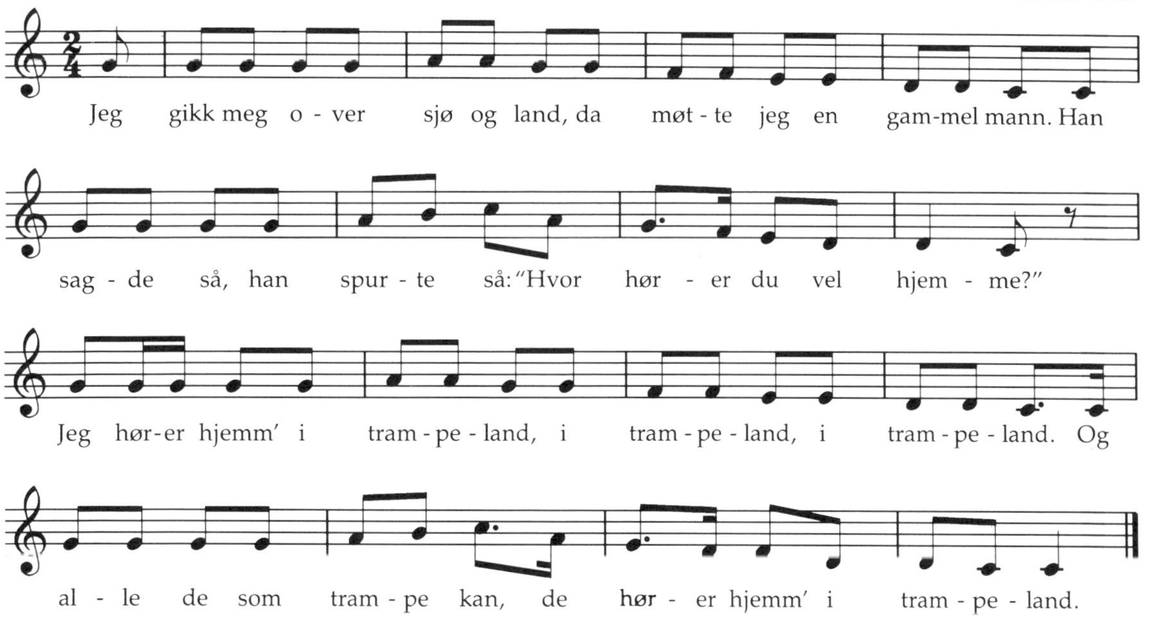

1. *Refrain:*
   Jeg gikk meg over sjø og land
   da møtte jeg en gammel mann.
   Han sagde så, han spurte så,
   "Hvor hører du vel hjemme?"

   * * *

   Jeg hører hjemm' i trampeland,
   i trampeland, i trampeland.
   Og alle de som trampe kan,
   de hører hjemm' i trampeland.

2. *Refrain*

   Jeg hører hjemm' i pekeland,
   i pekeland, i pekeland...

3. *Refrain*

   Jeg hører hjemm' i hinkeland,
   i hinkeland, i hinkeland...

4. *Refrain*

   Jeg hører hjemm' i hoppeland,
   i hoppeland, i hoppeland...

5. *Refrain*

   Jeg hører hjemm' i klappeland,
   i klappeland, i klappeland...

1. *Refrain:*
   I traveled over sea and land
   and then I met an old, old man.
   He said to me, he asked of me,
   "O where, o where can your home be?"

   * * *

   I belong in Stamping Land,
   in Stamping Land, in Stamping Land.
   And all of you who stamp, stamp can,
   you all belong in Stamping Land.

2. *Refrain*

   I belong in Pointing Land,
   in Pointing Land, in Pointing Land...

3. *Refrain*

   I belong in Limping Land,
   in Limping Land, in Limping Land...

4. *Refrain*

   I belong in Hopping Land,
   in Hopping Land, in Hopping Land

5. *Refrain*

   I belong in Clapping Land,
   in Clapping Land, in Clapping Land...

Å kjøre vatten

To Haul Water

Traditional

1. Å kjøre vatten, og kjøre ved,
og kjører tømmer over heia!
Og kjører å dom kjører vil,
je kjører jenta mi ei'a.

Refrain:
Di rau'e roser og di auer blå,
di vakre jenter holder je utå.
Helst når je får den je vil ha,
da er det morosamt å leva.

2. Det står et tre i min faders gård.
Det har så underli'e greiner;
Og visst je ikke blir gift i år,
så vil je leva aleine.

Refrain

1. Hauling water and hauling wood,
hauling lumber over the hill!
They can haul whatever they will –
I'll haul my own girl.

Refrain:
Those rosy cheeks and eyes of blue –
I like the beautiful girls.
And when I find the one I want,
then it will be fun to live.

2. There stands a tree in my father's yard.
It has such strange branches.
And if I don't get married this year,
then I want to live alone.

Refrain

3. Å kjører å dom kjøre vil,
je kjører kjerringa me ei'a!
Hu er så god, og hu er så snill,
je kan visst aldri bli lei'a.
Di rau'e roser og di øyne blå,
me ei'a kjerring holder je utå.
Nå har je fått, den je vil ha,
nå er det morosamt å leva!

3. They can haul whatever they like –
I'll haul my own wife!
She is so good, and she is so kind,
I will never get tired of her.
Those rosy cheeks and eyes of blue –
I like my own wife.
Now I have found the one I want,
and now it is fun to live!

Fram dansar ein haugkall

A Hill-Man Dances Forth

Olav Sande

1. Fram dansar ein haugkall fager og blå,
med gullring om håret som fløymer.
Han giljar for veslemøy til og frå
og tonar ikring honom strøymer.
"Å hildrande du, med meg skal du bu
i blåhaugen skal du din sylvrokk snu."

1. A hill-man dances forth, handsome and blue,
with gold rings in his flowing hair.
He entreats the young maiden all he can
and the melodies stream about her.
"Oh hey clever you, with me you shall live;
in the blue hill you will turn your
silver spinning wheel."

2. "Om dagen er eg den brune bjørn
som bysker i skogen vide
og laugar mitt ragg i djupaste tjørn
og vassar i straumane stride
og leikar i strand og rår over land
så langt som ditt auga det timja kan."

3. "Men når det lid til midnettes bel
og dagen i haug er gjengen
då høyrer du klunk og yndeleg spel
som linnaste lokking frå strengen;
då kjem eg til deg frå villande veg
og søv i din arm til det vekkjer meg."

4. "Men du skal sitje i blåhaug brur
i silke og sylv så det bragar
og aldri kjennast sår eller stur
i alle dei levedager.
Å hildrande du! Med meg skal du bu
i blåhaugen skal du din sylvrokk snu."

2. "During the day I am the brown bear
who bounds through wide forests
and bathes his fur in deepest lakes
and wades through swift rivers
and plays on the shores and rules over the lands
as far as your eyes can see."

3. "But when it approaches midnight time
and the day has passed over this hill,
then you will hear the elegant sound
of plucking strings that gently call;
then I will return to you from the wild ways
and sleep in your arms til I wake."

4. "But you shall reign as the blue-hill bride
with silks and silver to dazzle the eye
and never shall you know hurt or sadness
in all your living days.
Oh hey clever you, with me shall you live;
in the blue hill you will turn your
silver spinning wheel."

Den bakvende visa

The Backwards Song

Traditional

1. Eg beisla min støvel og sala mitt sverd,
så batt eg merra ved sida.
Så reiste eg meg til kattemordsferd–
skulle stande den farlege strida.

Refrain:

Eg låg og eg datt, eg drøymde i natt,
eg totte den visa var bakvend radt.

2. Og laksen høgt i furutopp,
han braut små lauvet av greiner.
Og ikorn ned på havsens botn
skulle brote opp store grå steinar.

Refrain

3. Og reven var raud og ekorn var snau,
det stod tvo dyr uti skogen.
Der stande tvo ikorn og tamde ein bjørn,
dei ville vel hava'n fyr plogen.

Refrain

4. Så kom det fram ein gamal blind mann,
skulle sjå kor det let no med tida.
Og stjernene song og månen han gol;
det skein ein gauk oppi lia.

Refrain

1. I bridled my boot and saddled my sword,
then tied the mare at my side.
I traveled away to battle the cats–
to stop that dangerous struggle.

Refrain:

I lay and I fell, I dreamed last night,
I thought this song was all backwards.

2. And the salmon high in the pine tree,
he broke small leaves off the branches.
And squirrels down at the bottom of the ocean,
were breaking up big grey stones.

Refrain

3. And the fox was red and the squirrels were bare,
there stood two animals out in the forest.
There are two squirrels taming a bear,
they probably want him to pull a plow.

Refrain

4. Then there came forth an old blind man,
–wanted to see what time it was.
And the stars sang and the moon called like a bird;
there shines a cuckoo over the hillside.

Refrain

Terningspelet

The Dice Game

Traditional

1. Prinsessa sat i høgan loft
med handi under kinn,
då kom ein fager ungersvein
gangande dit inn.

Refrain:
For dei leika, for dei leika,
dei leika og dei spela gullterning.

2. "Å høyr du fagre ungersvein
leik tavlebord med meg!"
"Eg heve inkje raudan gull
å setje opp mot deg."

Refrain

3. "Du setja inn din gode hatt,
enno om han er grå!"
"Så set og opp mi perlesnor,
freist om du ho kan få!"

Refrain

4. Då fyrste gong gullterningen
på tavlebordet rann,
Ungersveinen tapa han
og stolte jomfru vann!

Refrain

5. "Å høyr du gode ungersvein,
leik tavlebord med meg!"
"Eg heve inkje raudan gull
å setja opp mot deg."

Refrain

6. "Di kappe set du opp mot meg
med raude snorer på,
så set eg opp min ring av gull,
freist om du han kan få."

Refrain

7. Då andre gong gullterningen
kring tavlebordet rann,
ungersveinen tapa han
og stolte jomfru vann!

Refrain

8. "Å høyr du fagre ungersvein
leik tavlebord med meg!
Eg heve inkje raudan gull
å setje opp mot deg."

1. The princess sat in the high loft
resting her chin upon her hand.
Then a handsome peasant lad
came walking in.

Refrain:
And they played, and they played,
they played and gambled with dice of gold.

2. "Listen, my handsome peasant lad,
play at the game board with me!"
"I have no red gold
to bet against you."

Refrain

3. "You can wager your fine hat,
even if it is grey!
And I'll wager my string of pearls
– try to get them!"

Refrain

4. The first time the golden dice
ran across the game board,
the peasant lad lost
and the proud mainden won!

Refrain

5. "Listen, my good peasant lad,
play at the game board with me!"
"I have no red gold
to bet against you."

Refrain

6. "You can wager your cape
with the red braid.
And I'll wager my ring of gold
– try to get it"

Refrain

7. The second time the golden dice
ran across the game board,
the peasant lad lost
and the proud maiden won!

Refrain

8. "Listen, my handsome peasant lad,
play at the game board with me!"
"I have no red gold
to bet against you."

Refrain

9. "Du setja dine hosor inn
og dine ferdaskor,
så set eg opp alt det eg hev,
me ære og mi tru!"

Refrain

10. Då tredja gong gullterningen
kring tavlebordet rann,
stolte jomfru tapa ho,
og ungersvein han vann.

Refrain

11. "Høyr du gode ferdasvein,
du skunda deg bort frå meg.
Min kvite hest og dyre sal
så vil eg gjeva deg."

Refrain

12. "Din kvite hest og dyre sal
fær eg når eg kan.
Men eg vil ha den jomfru ven,
eg med gullterning vann!"

Refrain

13. Prinsessa sine hender vreid
og gret så mang ei tår.
"Gud trøyste meg for fattigsvein
som eg til brudgom får!

Refrain

14. "Nei, eg er ingen fattigsvein
om enn det tykkjest så.
Eg er den gjævast kongeson,
som soli skine på."

Refrain

15. Med krune av gull han rider no
rett fram til brudlaupsgård.
Med raude rosor flettar ho
sitt vene gule hår.

Refrain

Refrain

9. "You can wager your stockings
and your traveling shoes.
And I will wager all I have
and pledge myself as well."

Refrain

10. The third time the golden dice
ran across the game board
the proud maiden lost
and the peasant lad won.

Refrain

11. "Listen, good peasant lad,
hurry away from here –
My white horse and expensive saddle
will I give to you."

Refrain

12. "Your white horse and expensive saddle
I'll get when I can.
But I will have the fair maiden
who I won with the golden dice!"

Refrain

13. The princess wrung her hands
and cried so many a tear.
"God comfort me for that poor peasant lad
I've gotten for a husband."

Refrain

14. "No, I am not any poor peasant lad,
even thought one might think so.
I am the finest king's son
that the sun has shown upon."

Refrain

15. With a crown of gold, now he rides
forward to the wedding.
With red roses she braids
her lovely golden hair.

Refrain

Trolldans

Troll Dance

Traditional

1. Det leikar så mangt ei sumarnott,
   i dvergeskog, i dvergeskog.
   Alvedans og huldrelått –

   Refrain:
   Å hei deg å hildrande du!
   Lettlege trøda me dansen i kveld under fjelli!
   Lettlege trøda me dansen i kveld under himmeltjeld.

   Men sing:
2. "Å høyrer du, Gro kva eg seier deg,
   i dvergeskog, i dvergeskog.
   Vil du til Blåfjellet rida med meg?"

   Refrain

   Women sing:
3. "Å høyrer du, Per, kvi talar du sa?
   i dvergeskog, i dvergeskog.
   Fyrr skal det springe mitt hjarteblod."

   Refrain

1. There's dancing many a summer night,
   in the dwarves' forest, in the dwarves' forest.
   Elfin dance and huldre song –

   Refrain:
   Oh hei you oh hey – clever you!
   Lightly we step the dance tonight under the mountains!
   Lightly we step the dance tonight under the heaven's tapestry.

   Men sing:
2. "Listen here, Gro, to what I say to you,
   in the dwarves' forest, in the dwarves' forest.
   Will you ride to Blue Mountain with me?"

   Refrain

   Women sing:
3. "Listen here, Per, why do you speak like that?
   in the dwarves' forest, in the dwarves' forest.
   I would rather have my heart burst."

   Refrain

Men sing:

4. "Eg syrgjer no ikkje så såre for di,
i dvergeskog, i dvergeskog.
Heime ventar meg kjerringar tri."

Refrain

Men sing:

4. "I won't grieve much over you,
in the dwarves' forest, in the dwarves' forest.
Three wives are waiting at home for me."

Refrain

Eg gjekk meg opp til sæterli

I Went Up to the Mountain-Farm Meadow

Traditional

1. :/: Eg gjekk meg opp til sæterli
til jenta mi, ho var ikkje blid. :/:
Ho var ikkje grei, ho svara meg nei,
"Kva vil du no her, du fær ikkje meg."

2. :/: Men som det leid til laurdagskveld,
eg pynta meg og gjekk i veg lell. :/:
Og vegen var bratt, eg gjekk og eg datt,
eg kom 'kje fram før langt på natt.

3. :/: Men som eg kom til sæterstad,
då vart eg glad – ho svara meg ja. :/:
"Eg er ikkje vond, du skal få rom,
eg vart så glad eg såg du kom."

4. :/: Så hengd' ho over gryta si
med rømme 'ti og kokte til meg. :/:
Og grauten var heit, og han var feit,
og eg sku' skunda meg – det var 'kje
greitt.

1. :/: I went up to the mountain-farm meadow
to [see] my girl; she was not in good humor. :/:
She was not agreeable, she answered me, "No."
"What do you want here? You won't get me."

2. :/: But when Saturday evening came,
I got dressed up and walked up the road anyway. :/:
And that road was steep, I tripped and fell,
I didn't arrive until late in the evening.

3. :/: But once I got to the mountain-farm,
I was glad then – she answered me, "Yes." :/:
"I am not angry, there is a place for you,
I was so glad when I saw you coming."

4. :/: So she took out a pot
and cooked some porridge for me. :/:
And that porridge was hot, and it was rich,
and I shouldn't have eaten so fast – that
wasn't so good.

Eg gjætte Tulla

I Watched Over Tulla

Traditional

1. Eg gjætte Tulla i femten år
   eg passa vel på Tulla.
   Eg gjætte bort både lam og får,
   men endå hadde eg Tulla.

   Refrain:
   Å hei, å hå, det fær så gå!
   Eg tregar meste på Tulla
   for ho var krulla i ulla.

2. Eg gjætte Tulla bort i ei li,
   det var så lite til beite,
   eg gjekk og lokka på Tulla mi,
   og etter graset mun leite.

   Refrain

3. Og så kom skrubben laskande fram,
   då fekk eg kjøpmann til Tulla,
   og akkederingen var 'kje lang,
   og pengan' var ikkje mange.

   Refrain

4. Eg fekk ikkje meire av Tulla mi
   enn ein liten flekk utav blodet
   No kjem eg aldri til Fagerli,
   om der er gras nok og groe.

   Refrain

1. I had Tulla for fifteen years
   I took good care of Tulla
   I lost both lambs and sheep,
   but still I had Tulla.

   Refrain:
   Oh hei, oh ho, and so it goes,
   I grieve the most for Tulla,
   For she had such curly wool.

2. I had Tulla in a flock on a hillside
   There was so little to graze on.
   I went and called to my Tulla
   and looked for her in the grass.

   Refrain

3. And then a wolf went sneaking away.
   I found a buyer for what was Tulla
   and the agreement didn't take long
   and the money wasn't much.

   Refrain

4. I didn't have anything more of Tulla
   than a little bit of wool.
   Now I never go out to the beautiful hillside,
   even if there is fresh new grass.

   Refrain

Ola Glomstulen

Ola Glomstulen

Traditional

1. Ola Glomstulen hadde ei gamal grå geit.
   Ola Glomstulen hadde ei gamal grå geit.
   Å kjære mi Kari gjer pølsa vel feit!

   Refrain:
   For i morgon skal Ola Glomstulen gifta seg.

2. Ola Glomstulen hadde ei fingerbjørg malt.
   Ola Glomstulen hadde ei fingerbjørg malt.
   Å kjære mi Kari du brygger vel alt;

   Refrain

3. Ola Glomstulen hadde så lang ei brureferd.
   Ola Glomstulen hadde så lang ei brureferd.
   Ho nådde fra Solum til Gråten hjå Blehr;

   Refrain

1. Ola Glomstulen had an old grey goat.
   Ola Glomstulen had an old grey goat.
   Oh, Kari my dear, make that sausage good and fat!

   Refrain:
   For tomorrow Ola Glomstulen will get married.

2. Ola Glomstulen had a little malt.
   Ola Glomstulen had a little malt.
   Oh, Kari my dear, you had better brew it all.

   Refrain

3. Ola Glomstulen had such a long wedding procession.
   Ola Glomstulen had such a long wedding procession.
   It stretched from Solum to Gråten at Blehr's.

   Refrain

Ung Åslaug

Young Åslaug

Kristofer Janson

1. Ung Åslaug gjekk i bjørkeskog,
   så tirøygd, frisk og raud,
   ho tykte mest, at alle gret,
   der som ho gleda baud.
   "Kva er det vel dei syta for
   i verdi vid og stor?
   når soli skin, og hjarta brenn,
   og kjærleiken fær ord."

2. Då kom eit bod, at hennar mor
   låg sjuk og måtte døy,
   då kom eit bod, at hennar ven
   ha' fest ei onnor møy.
   Ung Åslaug vart så underleg,
   ho tagde reint den kveld,
   for noko bort frå hjarta fraus,
   endå det brann som eld.

1. Young Åslaug walked in the birch woods,
   so bright-eyed, cheerful, and rosy- cheeked.
   It seemed to her that everyone was sad,
   while she was full of happiness.
   "What is there to worry about
   in the world so wide,
   when the sun is shining, the heart is aflame,
   and love is in your thoughts?"

2. Then there came a message that her mother
   was sick and about to die;
   then there came a message that her lover
   was engaged to another maiden.
   Young Åslaug became so queer –
   she was very quiet that night,
   because something froze in her heart,
   even though it burned like a fire.

3. Ung Åslaug gjekk i bjørkeskog,
så tårøygd og så raud,
ho tykte mest, at alle lo,
der som ho sorgi baud.
"Kva er det vel dei fagnast ved
i verdi vid og stor?
når soli brenn, men hjarta frys,
og kjærleik bert er ord."

3. Young Åslaug walked in the birch woods,
so teary-eyed and red-faced.
It seemed to her that everyone laughed,
while she was full of sadness.
"What is there of worth
in the world so wide,
when the sun is burning, the heart is frozen,
and love is just a word?"

Eg rodde meg ut på seie grunnen

I Rowed Out to the Fishing Grounds

Traditional

Eg rodde meg ut på seie grunnen
det var om morgonen tidleg.
Då kom han Olav frå Kåremunnen
og lagde båten for ile.
Då dreiv eg til han med fiske stongi
så 'n datt ned i uvitet bak i rongi.
Eg vart så glad, tok til og kvad
Eg rådde grunnen åleine.
Sudeli, sudeli, sudeli dei – hoi!

I rowed out to the fishing grounds
It was early in the morning.
Then came that Olav from Kåremunnen
and anchored his boat.
So I hit him with my fishing pole,
and he fell senseless on the stern.
I was so glad, I began to sing.
I rowed to the fishing grounds alone.
Sudeli, sudeli, sudeli dei – hoi!

Hei huskom i hei!

Hey What A Hullabaloo

Traditional

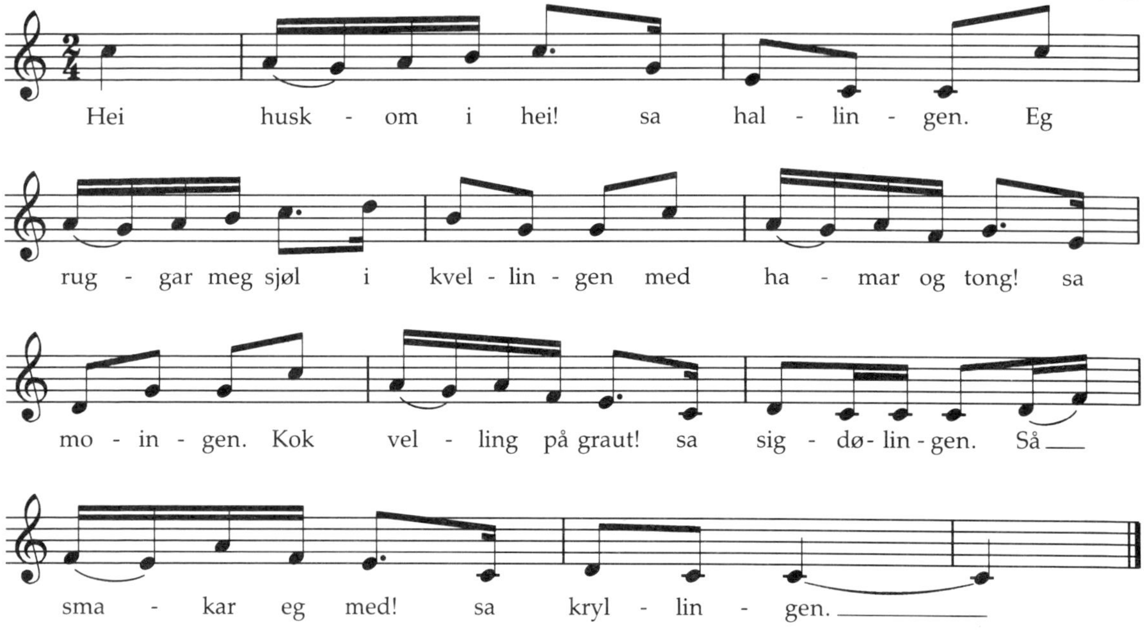

1. "Hei huskom i hei," sa hallingen.
"Eg ruggar meg sjøl i kvellingen
med hamar og tong," sa moingen.
"Kok velling på graut," sa sigdølingen.
"Så smakar eg med," sa kryllingen.

2. "Kor var du i natt?" sa jonsingen.
"Av og kjøpte meg ein katt," sa tentingen.
"Kva ville du med den?" sa jonsingen.
"Eg kjøpte'n for min ven," sa tentingen.
"Eg kjøpte'n for min ven," sa tentingen.

3. "Kor lever han Jo med båten sin?"
"Han ligg utpå havet og fiskar sild
og får han 'kje sild, så et han kling."
"Å silda var sur," sa hallingen.
"Å du talar som ein tjuv," sa noringen.

4. "Flink kjerring hev eg," sa Husåsen.
"To slike hev eg," sa Luråsen.
"Hot vi' du ha for ei," sa Husåsen.
"Ei alen med tobakk," sa Luråsen.
"Det var gampe verd," sa Husåsen.

1. "Hey what a hullabaloo," said the Hallingdal man.
"I lumber along in the evening with hammer and pliers," said the man from Mo.
"Cook porridge into gruel," said the Sigdal man.
"Then I'll taste it," said the Krødsherad man.

2. "Where were you tonight?" asked the Jøndal man.
"Off to buy me a cat," said the man from Tinn.
"What will you do with it?" asked the Jøndal man.
"I bought it for my friend," said the man from Tinn.
"I bought it for my friend," said the man from Tinn.

3. "Where does Jo live with his boat?"
"He is out on the ocean and fishes herring.
And if he gets no herring, then he eats lefse."
"Oh the herring was sour," said the Hallingdal man.
"Oh you talk like a thief," said the man from Nore.

4. "I have a clever wife," said Husåsen.
"Two such have I," said Luråsen.
"How much do you want for one?" asked Husåsen.
"Two feet of tobacco," said Luråsen.
"That is awful expensive," said Husåsen.

Familie vals

Seattle Polka

Ellingson's polka

Presented with permission of LeRoy Larson

Tretur fra Hordaland

Krossadans med seks

Hoppvals

Åttetur

Byte reinlender

Quickly

Figaro

Totur

Reinlender

Røros ringlender

Rugen

Firetur fra Landvik

Springpolka

Feiar fra Vestlandet

Seierstad Hopsa

Firetur fra Romerike

Klappdans

Lettisk Polka

Reinlender

Tretur fra Fana

Feiar med vals

Tantoli

Seksmannsril

Sunnmørsreinlender

Sekstur

Sandsværril

Parisarpolka

Record Shops with Norwegian Dance Music

Haug Forretning
2428 NW 56
Ballard, WA 98107
Phone (206) 789-3620

Norsk, Ltd.
800 Linden Avenue
Boulder, CO 80302
Phone (303) 442-6452

Skandisk Music
3424 19th Avenue South
Minneapolis, MN 55407
Phone (612) 722-1202

About the Author

Daniel Beal received a master's degree in Dance from the University of Oregon in 1978. In 1979, he was awarded a grant from the Norwegian government to conduct research on Norwegian ethnic dance and music. The grant made it possible for him to live and study in different parts of Norway over the course of one year. On this and subsequent trips, he gained a broad understanding of Norwegian folk culture. Some of his research has been presented in programs for public schools, lectures and workshops for universities in the Midwest, a radio documentary for Minnesota Public Radio, and an article for the *Journal of Ethnomusicology.* With his wife Gaelyn, the author has taught and performed Norwegian dances professionally in the United States and Canada for over 12 years. He has worked as a teacher and director of dance for international language camps, as well as an artist in residence for Minnesota public schools.